Time Telling through the Ages

HARRY CHASE BREARLEY

1919

TABLE OF CONTENTS

PREFACE
FOREWORD
THE MAN ANIMAL AND NATURE'S TIMEPIECES
THE LAND BETWEEN THE RIVERS
HOW MAN BEGAN TO MODEL AFTER NATURE
TELLING TIME BY THE WATER THIEF
HOW FATHER TIME GOT HIS HOUR GLASS
THE CLOCKS WHICH NAMED THEMSELVES
THE MODERN CLOCK AND ITS CREATORS
THE WATCH THAT WAS HATCHED FROM THE
NUREMBURG EGG
HOW A MECHANICAL TOY BECAME A
SCIENTIFIC TIMEPIECE
THE WORSHIPFUL COMPANY AND ENGLISH
WATCHMAKING
WHAT HAPPENED IN FRANCE AND
SWITZERLAND
HOW AN AMERICAN INDUSTRY CAME ON
HORSEBACK
AMERICA LEARNS TO MAKE WATCHES
CHECKERED HISTORY
THE WATCH THAT WOUND FOREVER
THE WATCH THAT MADE THE DOLLAR FAMOUS
PUTTING FIFTY MILLION WATCHES INTO
SERVICE
THE END OF THE JOURNEY
APPENDIX A
APPENDIX B
APPENDIX C
APPENDIX D
APPENDIX E

PREFACE

In the midst of the world war, when ordinary forms of celebration seemed unsuitable, this book was conceived by Robt. H. Ingersoll and Bro., as a fitting memento of the Twenty-fifth Anniversary of their entrance into the watch industry, and is offered as a contribution to horological art and science. Its publication was deferred until after the signing of the peace covenant.

The research work for fact material was performed with devoted fidelity and discrimination by Mrs. Katherine Morrissey Dodge, who consulted libraries, trade publications, horological schools and authorities in leading watch companies. The following were helpfully kind to her: New York Public Library, New York City; The Congressional Library, Washington, D. C.; Newark Public Library, Newark, New Jersey; The Jewelers' Circular, New York City; Keystone Publishing Company, Philadelphia, Pennsylvania; Mr. John J. Bowman, Lancaster, Pennsylvania; Major Paul M. Chamberlain, Chicago, Illinois; Hamilton Watch Company, Lancaster, Pennsylvania; Mr. Henry G. Abbott, of the Calculagraph Company, New York City, and others.

Credit is also due to Mr. Walter D. Teague, the well-known artist of New York City, who acted as art editor and supervised the preparation of illustrations, typography and other art and mechanical features.

The photographic compositions are the result of the enthusiasm, the understanding and the art of Mr. Lejaren a' Hiller, of New York City. In this connection the courtesy of Mr. Henry W. Kent, Secretary of the Metropolitan Museum of Art, New York City, in permitting the use of collections of the museum in the preparation of illustrations, is appreciated.

Harry C. Brearley

FOREWORD

It was a moonless night in No Man's Land. A man in khaki stood silently waiting in a frontline trench. In the darkness, his eyes were drawn, fascinated, to the luminous figures on the watch-dial at his wrist. A splinter of pale light, which he knew to be the hour-hand, rested upon the figure 11. A somewhat longer splinter crept steadily from the figure 12.

"Past eleven," he whispered to himself. "Less than twenty minutes now."

To the right and to the left of him, he, now and then, could see his waiting comrades in the blackness of the trench, their outlines vaguely appearing and disappearing with the intermittent flares of distant star-shells. He knew that they, too, were intent upon tiny figures in small luminous circles and upon the steady, relentless progress of other gleaming minute-hands which moved in absolute unison with the one upon his own wrist. He knew, also, that far in the rear, clustered about their guns, were other comrades tensely counting off the passing minutes.

At twenty minutes past eleven, the artillery bombardment would begin and would continue until exactly midnight. Then would come the barrage the protecting curtain of bursting shells behind which the khaki-clad figure and his companions would advance upon the enemy's trenches perhaps also upon eternity.

How strangely silent it seemed after the crashing chaos of the last few days! There were moments when the rumble of distant guns almost died away, and he could hear the faint ticking of his timepiece or a whispered word out of the darkness near at hand. He likened the silence to the lull before a storm.

Five minutes thus went by!

In another fifteen minutes, the fury of the bombardment would begin; it would doubtless draw an equally furious bombardment from the enemy's guns.

At twelve-ten plus forty-five seconds, he and his platoon were to "go over the top" and plunge into the inferno of No Man's Land. That was the moment set for the advance the moment when the barrage would lift and move forward.

The slender hand on the glowing dial stole steadily onward. It was ten minutes after now.

Ten minutes after eleven just one hour plus forty-five seconds to wait! His thoughts flew back to his home in the great city beyond the sea.

Ten minutes after eleven why that would be only ten minutes after six in New York! How plainly he could picture the familiar scenes of rushing, bustling life back there! Crowds were now pouring into the subways and surface cars or climbing to the level of the "L's." This was the third the latest homeward wave. The five o'clock people had, for the most part, already reached their homes and were thinking about their dinner; the five-thirties were well upon their way.

How the millions of his native city and of other cities and towns, and even of the country districts, all moved upon schedule! Clocks and watches told them when to get up, when to eat their breakfasts, when to catch their trains, reach their work, eat their lunches, and return to their homes. Newspapers came out at certain hours; mails were delivered at definite moments; stores and mills and factories all began their work at specified times.

What a tremendous activity there was, back there in America, and how smoothly it all ran smooth as clock-work! Why, you might almost say it ran by clock-work! The millions of watches in millions of pockets, the millions of clocks on millions of walls, all running steadily together these were what kept the complicated machinery of modern life from getting tangled and confused.

Yes; but what did people do before they had such timepieces? Back in the very beginning, before they had invented or manufactured anything far back in the days of the caveman even those people must have had some method of telling time.

A bright star drew above the shadowy outline of a hill. At first the man in khaki thought that it might be a distant star-shell; but no, it was too steady and too still. Ah yes, the stars were there, even in the very beginning and the moon and the sun, they were as regular then as now; perhaps these were the timepieces of his earliest ancestors.

A slight rustle of anticipation stirred through the waiting line and his thoughts flashed back to the present. His eyes fixed themselves again on the ghostly splinters of light at his wrist. The long hand had almost reached the figure 4 the moment when the bombardment would begin.

He and his comrades braced themselves and the night was shattered by the crash of artillery.

THE MAN ANIMAL AND NATURE'S TIMEPIECES

The story of the watch that you hold in your hand to-day began countless centuries ago, and is as long as the history of the human race. When our earliest ancestors, living in caves, noted the regular succession of day and night, and saw how the shadows changed regularly in length and direction as day grew on toward night, then was the first, faint, feeble germ of the beginning of time-reckoning and time-measurement. The world was very, very young, so far as man was concerned, when there occurred some such scene as this:

It is early morning. The soft, red sandstone cliffs are bathed in the golden glow of dawn. As the great sun climbs higher in the eastern sky, the sharply outlined shadow of the opposite cliff descends slowly along the western wall of the narrow canyon. A shaggy head appears from an opening, half-way up the cliff, and is followed by the grotesque, stooping figure of a long-armed man, hairy and nearly naked, save for a girdle of skins. He grasps a short, thick stick, to one end of which a sharpened stone has been bound by many crossing thongs, and, without a word, he makes his way down among the bushes and stones toward the bed of the creek.

Another head appears at the same opening in the cliff that of a brown-skinned woman with high cheek-bones, a flat nose, and tangled hair. She shouts after the retreating form of the man, and he stops, and turns abruptly. Then he points to the edge of the shadow far above his, and, with a sweeping gesture, indicates a large angular rock lying in the bed of the stream near by. Apparently understanding the woman nods and the man soon disappears into the brush.

The forenoon wears along, and the line of shadow creeps down the face of the canyon wall until it falls at last across the angular rock against which the dashing waters of the stream are breaking. The woman who has been moving about near the cave opening begins to look expectant and to cast quick glances up and down the canyon. Presently the rattle of stones caught her ear and she sees the long-armed man picking his way down a steep trail. He still carries his stone-headed club in one hand, while from the other there swings by the tail the body of a small, furry animal. Her eyes flash hungrily, and she shows her strong, white teeth in a grin of anticipation.

"I'll be back when the shadow touches that stone." It was by such crude expedients that our primitive ancestors timed their engagements.

Perhaps it has not been hard to follow the meaning of this little drama of primitive human need. Our own needs are not so very different, even in this day, although our manners and methods have somewhat changed since the time of the caveman. Like ourselves, this savage pair awoke with sharpened appetite, but, unlike ourselves, they had neither pantry nor grocery store to supply them. Their meal-to-be, which was looking for its own breakfast among the rocks and trees, must be found and killed for the superior needs of mankind, and the hungry woman had called after her mate in order to learn when he expected to return.

No timepieces were available, but that great timepiece of nature, the sun, by which we still test the accuracy of our clocks and watches, and a shadow falling upon a certain stone, served the need of this primitive cave-dweller in making and keeping an appointment.

The sun has been, from the earliest days, the master of Time. He answered the caveman's purpose very well. The rising of the sun meant that it was time to get up; his setting brought darkness and the time to go to sleep. It was a simple system, but, then, society in those days was simple and strenuous.

For example, it was necessary to procure a new supply of food nearly every day, as prehistoric man knew little of preserving methods. Procuring food was not so easy as one might think. It meant long and crafty hunts for game, and journeys in search of fruits and nuts. All this required daylight. By night-time the caveman was ready enough to crawl into his rock-home and sleep until the sun and his clamoring appetite called him forth once more. In fact, his life was very like that of the beasts and the birds.

But, of course, he was a man, after all. This means that a human brain was slowly developing behind his sloping forehead, and he could not stop progressing.

After a while a long while, probably we find him and his fellows gathered together into tribes and fighting over the possession of hunting-grounds or what not, after the amiable human fashion. Thus, society was born, and with it, organization. Tribal warfare implied working together; working together required planning ahead and making appointments; making appointments demanded the making of them by something by some kind of a timepiece that could indicate more than a single day, since the daily position of light and shadows was now no longer sufficient. Man looked to the sky again and found such a timepiece.

Next to the sun, the moon is the most conspicuous of the heavenly objects. Its name means "the Measurer of Time." As our first ancestors perceived, the moon seemed to have the strange property of changing shape; sometimes it was a brilliant disk; sometimes a crescent; sometimes it failed to appear at all. These changes occurred over and over again always in the same order, and the same number of days apart. What, then, could be more convenient than for the men inhabiting neighboring valleys to agree to meet at a certain spot, with arms and with several days' provisions, at the time of the next full moon? moonlight being also propitious for a night attack.

For this and other reasons, the moon was added to the sun as a human timepiece, and man began to show his mental resources he was able to plan ahead. Note, however, that he was not concerned with measuring the passage of time, but merely with fixing upon a future date; it was not a question of how long but of when.

This presumptuous, two-legged fighting animal, from whom we are descended, and many of whose instincts we still retain, began to enlarge his warfare, and thereby to improve his organization. For the sake of his own safety, he learned to combine with his fellows, finding strength in numbers, like the wolves in the pack; or, like ants and bees, finding in the combined efforts of many a means of gaining for each individual more food and better shelter than he could win for himself alone.

For example, it was possible that a neighboring tribe, instead of waiting to be attacked, was planning an attack upon its own account. It would not do to be surprised at night. Sentries must be established to keep watch while others slept, and to waken their comrades in case of need. Our very word "watch" is derived from the old Anglo-Saxon word "waeccan," meaning "wake." And yet people who tried to watch for long at a stretch would be apt to doze. They must be relieved at regular times; it was a matter of necessity, but how could one measure time at night?

Where man has been confronted with a pressing problem he has generally found its solution. Probably in this case the stars gave him a clue. If the sky were clear, their positions would help to divide the night into "watches" of convenient length.

Thus did primitive man begin to study the skies. No longer a mere animal, he was beginning, quite unconsciously, to give indications of becoming a student.

THE LAND BETWEEN THE RIVERS

Now we must jump over ages so vast in duration that all of our recorded history is by comparison, the merest fragment of time. During the prehistoric period, known to us only by certain bones, drawings, and traces of tombs and dwellings, and by a few rude implements, weapons, and ornaments, we must think of the human family as developing very, very slowly groping in the dawn of civilization while it ate and slept, hunted, and fought, and, gradually spread over various regions of the earth.

It was in this interval, also, that man learned the use of fire and the fashioning of various tools. His club gave place to the spear, the knife, and the arrow-head weapons that were made at first by chipping flakes of flint to a sharp edge. Then, as his knowledge and skill slowly increased, he learned to work the softer metals and made his weapons and his tools of bronze. Meanwhile, he was taught, by observing in nature, to tame and to breed animals for his food and use, and to plant near home what crops he wished to reap, instead of seeking them where they grew in a wild state. Thus, he became a herdsman and farmer.

He no longer lived in caves or rude huts, but in a low, flat-roofed house built of heavy, rough stone, and, later, of stones hewn into shape or of bricks baked in the burning sunshine. Stone and clay carved or molded into images, and the colored earth, smeared into designs upon his walls, gave him the beginnings of art. And from drawing rude pictures of simple objects, as a child begins to draw even before knowing what it means to write, primitive man came at last to the greatest power of all the art of writing.

Through all this age man continued to regulate his expanding affairs by the timepieces of the sky the sun, the moon, and the stars. He divided time roughly into days and parts of days, into nights and watches of the night, into moons and seasons determining the latter probably by the migration of birds, the budding of trees and flowers, the falling of leaves and other happenings in nature. But never guessing how greatly interested future generations would be in the way he did things, he has left only a few records of his activities and these have been preserved by the merest accident. The historian and the press-agent were the inventions of later days.

Thus we come down the ages to a date about 4000 B. C. at the very beginning of recorded history, and to one of the most ancient civilizations in the world that of the region which we now call Mesopotamia. Mesopotamia lies in southwestern Asia between the Tigris and Euphrates Rivers and not far from the traditional site of the Garden of Eden. The name by which we know it comes from the Greek, and means, "The land between the rivers" but the people who dwelt there at the time to which we refer called it the "Land of Shinar."

This is the region in which long afterward so the Bible tells us Abraham left his native town, Ur of the Chaldees, to make his pioneer journey to Palestine. This is the land where the great cities of Babylon and Nineveh afterward arose; Babylon, where Daniel interpreted the dream of King Nebuchadnezzar, and Nineveh, whence the Assyrians, the fierce conquerors of the ancient world, "came down like a wolf on the fold" against the peaceful Kingdom of Judah. It is the land where, thousands of years later, the famous Arab capital of Bagdad was built; it is the land of Harun al Raschid and the "Arabian Nights," and the land which the British Army conquered in a remarkable campaign against the Turks and Germans. Mesopotamia is a land of color, brilliant life, wonders and romance. Many students and statesmen believe that it will, in days to come, grow fruitful and populous again, that it will once more be great among the countries of the earth. It is a flat region, with wide-stretching plains. For the most part, there are no hills to limit the view of the skies, and the heavens are brilliant upon starry nights.

In this favored portion of the earth, a high civilization had already been developed in the very earliest days of which we have authentic historic record. The caveman type had long disappeared and had been forgotten; people were already living in well-built cities of brick and stone. Their houses were low and flat-roofed, but the cities were surrounded with high and massive walls to protect them from enemies, and here and there within rose great square towers which were also temples. Perhaps the famous Tower of Babel was one of these, for Babel, of course, is another name for Babylon, and its people are known to have worshipped on the tops of towers, as if, by so doing, they could reach nearer to their gods. The ancient Chaldeans were religious by nature, and because the skies contained the greatest things of which they knew, they identified many of their gods with the sun, the moon, and the stars, and they worshipped these in their temples.

Thus, the sun was the god Shamash, the moon was Sin, Jupiter was Marduk, Venus was Ishtar, Mars was Nergal, Mercury was Nebo, and Saturn was Ninib.

In consequence, their priests came to give much of their time to a study of the movements of the stars. These priests, who were shrewd and learned men, discovered a great deal, but they kept their knowledge closely within the circle of their caste. Learning was not for everyone in those days because the priests posed as magicians able to interpret dreams, to explain signs, and to foretell the future. This brought them much revenue; as prophets they were not unmindful of profits.

When we consider that these astrologer-astronomers did not have telescopes or our other modern instruments, it is marvelous to see how many of the laws of the heavenly bodies they really did find out for themselves. Books could be filled, with the story of their discoveries. For example, they observed that the sun slowly changed the points at which it rose and set. During certain months, the place of sunrise traveled northward, and at the same time the sun rose higher in the sky, and at noon was more nearly overhead. At this time, the days were also longer, because the sun was above the horizon more of the time, and then it was summer. During certain other months, the sun traveled south again, and all these conditions were reversed; the days grew shorter and shorter, and it was winter. This is, of course, exactly what the sun appears to do here and now, and we may observe it for ourselves. But these Babylonian priests were the first to study these phenomena and accomplish something by applying their reasoning powers to the facts that presented themselves. They took the time which was consumed in this motion from the furthest north to the furthest south and return, and from that worked out their year.

In order to calculate time, they next devised the zodiac, a sort of belt encircling the heavens and showing the course of the sun, and the location of twelve constellations, or groups of stars, through which he would be seen to pass if his light did not blot out theirs. They divided the region of these twelve constellations into the same number of equal parts; consequently, the sun passing from any given point around the heavens to the same point, occupied in so doing an amount of time that was arbitrarily divided into twelfths.

But they also devised another twelve-part division of the year. They noticed that the moon went through her phases, from full moon to full moon in about thirty days. So one moon, or one month, corresponded with the passage of the sun through one "sign" of the zodiac. Our own word "month" might have been written "moonth," since that is its meaning. That gave them a year of twelve months, each month having thirty days, or three hundred and sixty days in all.

Then from the seven heavenly bodies which they had identified with seven great gods, they got the idea of a week of seven days, one day for the special worship of each god and named for him.

In like manner, they divided the day and the night each into twelve hours; and the hour into sixty minutes and these again into sixty seconds. The choice of "sixty" was not a chance shot or accident; it was carefully selected for practical reasons since these old astronomers were wise and level-headed men. No lower number can be divided by so many other numbers as can sixty. Just look at your watch for a moment and notice how simply and naturally the minutes, divided into fives, fit into place between the figures for the hours, and, because sixty divides evenly by fifteen and thirty, we have quarter-hours and half-hours.

Therefore, we should realize, with a bit of gratitude, that we owe these divisions of time, of which we still make use, to the ancient magician-priests of Babylon and Chaldea, thousands and thousands of years ago.

In doing all this, these early scientists developed at the same time an elaborate system of so-called "magic" by which they pretended to foretell future events and the destinies of men born on certain days. This was an important part of their priestcraft, and probably it was not the least profitable part. In fact, the priests called themselves magi, meaning "wise men" in their language, and our word "magic" is derived from "magi."

This magic, or prophetic study of the stars, we call astrology to distinguish it from the true science of astronomy. But mingled with it all, these priests possessed a wonderful amount of genuine scientific knowledge. Their year of three hundred and sixty days was, of course, five days too short, as they presently found out for themselves. In six years, the difference would amount to thirty days, which was exactly the length of one of their months. So they corrected the calendar very easily by doubling the month Adar once in six years. Thus, every sixth year contained thirteen months instead of twelve; that was the origin of the leap-year principle which we still use, although more accurately. It can be seen that, with all their superstition and their befooling of other people, the priests themselves were by no means ignorant; they were really keen observers.

This calendar, by which we still measure the years and the seasons, is so interesting a thing that it is worth while to pause for a moment in our story in order to trace out its later development. The Babylonian calendar remained practically the same up to the time of Julius Caesar, only a few years before the Christian Epoch. The names of the months had naturally been changed into the Latin language; and the Romans, instead of doubling a whole month, had come to add the extra five days to several months, one day to each. That is the reason for some of our months having thirty-one days.

When Caesar was Dictator of Rome, it had become known that the year of exactly 365 days was still a little too short. It should have been 365¼. So Caesar in reforming the calendar, provided that the first, third, fifth, seventh, ninth, and eleventh months should be given thirty-one days each, and that the others should have thirty days, except in the case of February which should have its thirtieth day only once in four years. A little later, his successor, the Emperor Augustus, after whom the month of August is named, decided that his month must be as long as July, which was Julius Caesar's month. Therefore, he stole a day from February and added one to August; then he changed the following months by making September and November thirty-day months and giving thirty-one days to October and December.

The Julian calendar, with these changes by Augustus, remained in use until the year A. D. 1582, nearly a century after the discovery of America. Then it was learned that the average year of 365¼ days was still not exactly right according to the motion of the earth around the sun. The exact time is 365 days, 5 hours, 48 minutes and 46 seconds, being 11 minutes and 14 seconds less than 365¼ days. When, therefore, we add a day to the year every four years, as Caesar commanded, we are really adding too much. This excess was corrected by Pope Gregory XII in 1582, when he changed the calendar so that the last year of a century should be a leap-year only when its number could be divided evenly by 400. Thus, 1700, 1800, and 1900 were not leap-years, though the year 2000 will be. This new calendar, which is the one now generally in use in most of the world, is known as the Gregorian calendar.

Thus the plan and principle of the calendar, as well as our smaller divisions of time, in spite of the small changes by Caesar and Gregory, have remained from the Babylonian days down to the present; and we have done nothing to their system in all these thousands of years, except, incidentally to correct it.

Only once in history have the measures of the ancient calendar been set aside. That was in France at the time of the Revolution, when the French people, in their passionate hatred of all the traditional things that reminded them of their past sufferings, invented a new calendar, in which they changed the names of months and days, and counted the years from 1792, the first of their liberty. They also abolished all Sundays and religious festivals, and divided the day into ten hours. This played havoc with time-keeping, and caused great confusion. Watches and clocks were made with one circle of numbers for the new hours, and another, within, on which were shown the old hours which people could understand. But this complication lasted only a few years, for the traditional system was soon restored.

To return again to the era of the first calendar. While the wise men of

Mesopotamia were engaged in mingling science and mystery, another civilization, the Egyptian, was developing upon the banks of the Nile and passing through much of the same stages. In due course the Persians conquered both Mesopotamia and Egypt and absorbed their knowledge. Still later the wonderful Greek nation combined astronomy with mathematics in a way which makes us wonder to this day. This is the way in which civilization has grown. Race after race, during century after century has added its new knowledge and discoveries to that which has been learned before. It is interesting to note that the astronomy of the Babylonians appears to have been paralleled independently by other ancient civilizations between which there was no apparent possibility of intercourse. The Chinese in the East and the Aztecs of Mexico, on the other side of the world, invented practically the same astronomical instruments as the Babylonians and made similar discoveries. All methods of indicating time have been steps upon the long road which has led to the making of modern timepieces.

The progressive Greeks did not permit knowledge to be monopolized by the priesthood and probably their common people knew more about the stars than most of the population of America do to this day. Sailors possessed no compasses, but they voyaged very skilfully with the guidance of the stars, while farmers, lacking our modern weather-reports and crop-bulletins, learned to govern their planting and harvesting by the positions of the heavenly bodies.

In one sense, this is time-telling and in another it is not, but our ideas of time and astronomy have always been so closely associated that it is hard to think of one apart from the other. This is because the movements of the earth, which produce night and day and the changes of the seasons, are our supreme court of time, our final standard for its measurement. And since we cannot see the earth move, we judge of its motion by the apparent movement of the heavenly bodies, just as we realize the movement of a train by watching the landscape rush past us as we go.

Some of the great Greek scientists, by the way, had even learned to foretell eclipses of the sun. According to Herodotus the one which occurred on May 28th, in the year 585 B. C., was predicted by Thales of Miletus, one of the famous "Seven Wise Men." This event was also celebrated because of another interesting association; it stopped a battle between the armies of the Medes and the Lydians. Perhaps we can guess at what happened. Undoubtedly the eclipse was interpreted by the armies as a sign of divine anger, for the ancients identified many of the forces and objects of nature as gods, and Phoebus Apollo, who it was believed daily drove his flaming chariot across the sky, was the great divinity of the sun. Furthermore, these gods were very apt to meddle with happenings upon the earth, particularly with wars, as anyone who has read the "Iliad" will recall.

The Chaldean priests in ancient Mesopotamia told time by the stars, thus combining science with religion.

Imagine, then, the two armies about to go to battle when suddenly something appeared to go wrong with the sun. There to their amazement, in a cloudless sky, a dimming shadow touched the edge of the sun's shining disk and began slowly to blot it out. The warriors forgot to fight each other and stared in terror at the sky. The sun dwindled to a crescent; a weird twilight fell upon the earth. Finally, the last thread of brightness disappeared leaving a dull circle in the sky, surrounded by faint bands of light. The gloom of night fell upon the ground. Birds and animals went to their rest.

No further evidence was needed by the superstitious and frightened soldiers. It must be true that Phoebus Apollo was grievously angered, and they forthwith laid down their arms. The sun god, of course, soon showed his approval of this action by coming back into the sky.

This is only one of many tales which might be told to show the state of superstition in those days. Learning, then, was confined to the few, and in many instances was used to mystify or terrorize the mass of the people and thus keep them submissive. At best, new ideas were slow to grow or to be believed.

For example, Pythagorus, the great Greek philosopher of the sixth century B. C., believed the earth to be a globe, but it was not until Columbus discovered America twenty centuries later that people generally began to know that it was not flat. Even in these modern days of the public school, the press, the telephone, the telegraph, the wireless and other means for the wide-spread distribution of knowledge, how slowly does truth find its way to acceptance! To this day, superstition is by no means dead.

Even Mark Twain, who scoffed at superstition all his life, often said that, as he came into the world with Halley's Comet, in the year 1835, so he expected to die in 1910, the year of the comet's next appearance. Strangely enough, his half-jesting prophecy was fulfilled, for he really did die in that year.

Astronomers to-day can figure out in advance what is to happen in the heavens with an exactness which would have seemed magical in olden times, and is hardly less astonishing even now. Their power is largely due to improved scientific instruments, proficiency in mathematics and greater accuracy in the measurement of time. Not only is the date of an eclipse of the sun now known in advance, but so also is the exact path of the shadow across the world, and the instant of its appearance in any given place.

We now have glanced briefly at a few of the features of early humanity's dependence upon the clocks of nature and the way in which they influenced its manner of life. We still depend upon these great primeval timepieces and we do it for the most part unconsciously, for our master clocks must still be set by the motion of the heavenly bodies.

That motion, which now we know to be really the revolution of our earth, is still the legislator and supreme court of time. But we have learned to make and carry everywhere a wonderful machine, whose revolving wheels and pointing hands keep tryst with the stars in the heavens and move to the rhythm of wheeling worlds. And so familiar is this talisman of man's making, that we forget to look beyond it or think of time at all save as the position of the hands upon the dial.

We carry with us carelessly a toy which tells tales upon the solar system our watch is a pocket universe.

HOW MAN BEGAN TO MODEL AFTER NATURE

We now have reached a point far ahead of our story and must take a backward step. We have been seeing man as a mere observer of nature; but man doesn't stop with nature as he finds it his man-brain drives him forward; he must make improvements of his own. Animals may live and die and leave no trace save their bones, which for the most part soon disappear, but man always leaves traces behind him. He has always interfered with nature, or rather has modeled after nature, seeing in her work the revelations of principles and laws that he might utilize in varying ways for his own benefit and progress. Our material civilization is built up from the accumulated results of all this study and control of nature by hundreds of millions of busy brains and hands, through tens of thousands of years.

Here we are, then, living, in a sense on the top of the ages of human history, like the dwellers on a coral island. Hundreds of generations have toiled to raise the vast structure for us, like the little coral "polyps" which build their own lives into the mass, yet we take it all as a matter of course and rarely give a thought to the marvelous ways by which it has come about. You may have just glanced at your watch. To you, perhaps, a watch has always seemed merely a small mechanism which was bought in a store. That is true, and yet remember this the first manufacturer who had a hand in producing that watch for you, may have been a caveman.

In order to appreciate this development, let us return, therefore, for another rapid view of prehistoric times; life in its crudest form one day much like another a scanty population, huddled in little groups in places naturally sheltered the simplest physical needs to be provided for little thought of the past or care for the future time-reckoning reduced to the single thought of appointment no reason for measuring intervals in these and other respects antiquity presented the greatest possible contrast to our complicated modern life.

The long-armed man of our first chapter noticed that as the sun moved, the shadows of the cliff also moved, as did all other shadows. As he formed habits of regularity, it was natural for him to perform a certain daily act when, perhaps, the shadow of a certain tree touched upon a certain stone. This would be a natural sun-dial.

But a thinner, sharper shadow would be easier to observe; suppose, therefore, that some successor to the long-armed man set up a pole in some open space and laid a stone to mark the spot where the shadow fell when the sun was highest in the heavens. That would be an artificial sun-dial a device deliberately planned to accomplish a certain purpose. The man who first took such a step was probably the first manufacturer who had a hand in supplying you with your watch. The shaggy mammoth, the terrible saber-tooth tiger and the eohippus, the small ancestor of our modern horse, must have been familiar sights when time-recording at the hands of some rude, unconscious inventor thus began the long story of its development.

One stone reached by the moving shadow would mark only one point of time each day. Why not place two stones, three stones, or even more and get more markings? Such a procedure would be more useful because it would indicate the time of other happenings in the course of the day. The sun would pass across the skies and the shadow must travel around the pole. What more natural than to place the stones in a circle and get a series of these markings?

Of course, as the ages passed, life became more complex not complex as we would consider it to-day, but, as compared with its rude beginnings. New habits were formed, new needs developed, new activities were undertaken at different periods.

Here, then, was the sprouting of modern civilization the beginning of that specializing of each man in his own particular direction that has carried the world to its present high state of expertness in so many fields. Slowly steadily, and inevitably this principle of specialization has been developed. With the increase of laws, for example, certain men came to give them special study and then to sell their knowledge and skill to other men who had no opportunity for such study. In course of time, the aggregation of laws became so great that these lawyers were forced to specialize among themselves; to-day, therefore, we find a number of classes of law specialists. The same thing is true of doctors who have limited their practise until we find those who treat the eye only, or the lungs, the stomach, or the teeth. Even the treatment of the teeth has been subdivided, some dentists limiting themselves to extraction and some of them even to the treatment of a single disease of the gums.

Engineering, too, has branched like a tree and the branches have branched again and yet again. Electrical engineering has come to be divided into so many departments that telephone companies employ specialists in many branches of the engineering profession.

We find the same conditions in any field of thought or activity all commercial and industrial life is divided and subdivided; labor is specialized; writing is specialized; teaching is specialized; even warfare has become a contest between many kinds of trained specialists, each employing the tools of his trade; and every man's outlook upon life is directed chiefly toward the particular corner of the particular field that he has fitted himself to occupy.

The first step toward this complex condition of the modern world was taken when each man stopped getting his own food, making his own weapons, and providing for all his individual wants without dependence upon others. When he learned to exchange that which he could best produce for that which some other man had learned to make better than he, the human race unconsciously turned away from the status of the birds and the beasts and began the long, slow upward climb that history records.

It was, then, through trade, barter and exchange that man began to acquire the manners of civilized life. Trade itself became a specialized activity, and dealers who did nothing but buy and sell, but themselves produced no material goods, found that a special calling was rightfully theirs. The modern merchant is the heir of one of the first "specialists" in human activity, and the misunderstood work of the so-called "middleman" is one of the bases of modern civilization a necessary and honorable calling.

The "Dial of Ahaz" was probably a flight of curving steps upon which a beam of sunlight fell. See Isaiah, xxxviii.

Civilization is a thing of the spirit, but it has the support of material things and it has been truly said that the degree of a people's civilization can be measured by the multiplicity of its needs. The savage is content with food, shelter and a covering for his body, but every step in civilization's progress has a more and more complex material accompaniment, and these interwoven relationships of modern life in which the question of time is a most important factor can only be sustained through the use of accurate time-measure. In other words, modern civilization leans upon the watch.

But here again we have run somewhat ahead of our story which, as a matter of fact, had only reached the point of primitive sun-dials. But this anticipation will be excused because of the importance of emphasizing that the growing interdependence of human relations had made it necessary to take into account the convenience of a greater and greater number of people, and this involved closer and closer time-recording in smaller divisions of time by more exact methods.

The sun-dial underwent so many changes that a volume would be needed to describe them all. For example, it was found that the shadow of an upright stick or stone varied from day to day, because, as we have already noticed, the sun rises farther north in summer in the northern hemisphere than it does in winter. So the mark for a certain hour would change as the season changed, and the dial would not indicate time accurately.

Berosus, a Chaldean historian and priest of Bel, or Baal, a god of the old Babylonian, lived about the year 250 B. C., and hit upon a very ingenious way of solving this difficulty. He made the dial hollow like the inside of a bowl. Into this the shadow was cast by a little round ball or bead at the end of a pointer that stood horizontally out over the bowl.

Now the sky itself is like a great bowl or inverted hemisphere, and, howsoever the sun moved upon it, the shadow would move in the same way upon the inside of the bowl or hemisphere. And by drawing lines in the bowl, similar to the lines of longitude upon the map, the hours could be correctly measured. The "Hemicycle of Berosus," as it was called, remained in use for centuries and was the favorite form of sun-dial all through the classic period of Greece and Rome. Cicero had one at his villa near Tusculum, and one was found, in 1762, at Pompeii.

But the hemicycle was not easy to make unless it were fairly small, and, if small, it was not very easy to read. You can see that a shadow which traveled only a few inches in a whole day would move so slowly that one could hardly see it go. And the shadow of a round ball is not a clear sharp-pointed thing like the hand of a watch, whose exact position can be seen however small it may be. Besides, the ancients were not very particular about exact timekeeping. They had no trains to catch, and in their leisurely lives convenience counted for more than doing things "on the minute." So they still continued using the upright pointer which the Greeks called the gnomon, meaning "the one who knows."

"Cleopatra's Needle," and other Egyptian obelisks may also have been used as huge gnomons to cast their shadows upon mammoth dials, for they were dedicated to the sun. With an object of such great size the shadow would move rapidly enough to be followed easily by the eye. But of course its motion would be irregular because of the flat surface of the dial. The word "dial," by the way, comes from the Latin dies meaning "day," because it determined the divisions of the day.

Then there was applied the idea of making the shadow move over a hollow space, such as a walled courtyard, going down one side, across, and up the other side as the sun went up, across and down the sky. Sometimes light was used instead of shadow, the place being partially roofed over and a single beam of light being admitted through a small hole at the southern end. Men kept track of the motion of this beam as it touched one point after another during the day.

Do you remember the miracle of the dial of Ahaz, mentioned in the Bible? Hezekiah the king was sick and despondent, and would not believe that he could ever recover from his illness or prevail against his enemies. So the prophet, Isaiah, in an effort to comfort the royal sufferer, made the shadow return backward ten degrees upon the dial of Ahaz, as a sign from heaven that his prophecy of the king's future recovery was true. You will find the story in Isaiah, Chapter thirty-eight.

This dial of Ahaz was probably a curved flight of steps rising like the side of a huge bowl at one end of the palace courtyard, with either a shadow cast by a pointer overhead or a beam of light admitted through an opening. It can be seen that this and similar great dials were applications of the hemicycle idea on a large scale.

According to our chronology, the dial of Ahaz must have been built during the eighth century, B. C. Although the sun-dial period was, of course, many hundreds of years older than this, yet the story of this Hebrew king and prophet is the first authentic reference to a sun-dial which has been discovered.

However, the final improvement of the dial was made when it was discovered that by slanting the pointer, or gnomon, exactly toward the north pole of the sky the point where the north star appears at night the sun's shadow could be cast upon a flat surface with accurate results in indicating time.

This may sound simple, but if you will look at a sun-dial such as may still be found in gardens, you will see that the lines of the hours and minutes are laid out on certain carefully calculated angles; you will realize that people had to acquire considerable knowledge before they were capable of making such calculations. The whole subject of dial-making is so complicated that, in 1612, there was published a big book of eight hundred pages on the subject.

The angles of the lines of the sun-dial must be different for different latitudes. It took that strong-arm race of ancient times, the Romans, a hundred years to learn this fact. The Romans, at this time, were developing their civilization from the shoulders downward, while the Greeks and some of the Greek colonies developed theirs from the shoulders upward. Rome was a burly power, with powerful military muscles. Whatever it wanted it went out and took at the point of the sword, as some nations have endeavored to do in latter days. Thus, the city of Rome became a vast storehouse of miscellaneous loot the fruit of other men's brains and hands.

Some conqueror of that day took back with him a sun-dial from the Greek colony of Sicily. This was set up in Rome, where nobody realized that even the power of Rome's armies was not able to transplant the angle of the sun as it shone upon Sicily far to the southward. It was nearly one hundred years before these self-satisfied robbers found that they had been getting the wrong time-record from the stolen instrument. Thus, the original owners had a form of belated revenge, could they but have known it.

One of the largest of all the sun-dials was the one set up by the Roman Emperor Augustus when he returned from his Egyptian wars bringing with him an obelisk not unlike the one which now stands near the Metropolitan Museum of Art in Central Park, New York City. If you can imagine this Egyptian obelisk, with its strange hieroglyphic characters upon its four sides, surrounded by a great dial with the figures of the hours marked upon its surface, you will get an idea of the size of this huge timepiece. However, it was probably more picturesque than valuable as a time-keeper.

There is an important difference between clocks and sun-dials, aside from the self-evident one of the difference in their construction. Clock-time is based on what is called "mean time." If we study the almanac table of times of sunrises and sunsets, and count the number of hours from sunrise of one day to sunrise of the next, we find it is rarely exactly twenty-four hours, but usually a few minutes more or less, while the average for the whole year is twenty-four hours. The clock is constructed to keep uniform time based on this average length of day.

The sun-dial time marks "apparent time," the actual varying length of each day. The sun-dial time, therefore, is nearly always some minutes ahead or behind that of a clock, the greatest discrepancy being about sixteen minutes for a few days in November. There are, however, four days in the year when the clock and the sun-dial agree perfectly in the time they indicate. These days are April 15th, June 15th, September 1st, and December 24th.

When in the eighteenth century clocks and watches began to come into wide-spread use sun-dials fell into neglect, except as an appropriate bit of ornament in gardens. At Castletown, in the Isle of Man, is a remarkable sun-dial with thirteen faces, dating from 1720.

It was usual to place on sun-dials appropriate mottoes expressing a sentiment exciting inspiration or giving a warning to better living. A dial that used to be at Paul's Cross, London, bore an inscription in Latin, which translated means, "I count none but the sunny hours." In an old sweet-scented garden in Sussex was a sun-dial with a plate bearing four mottoes, each for its own season: "After darkness, light;" "Alas, how swift;" "I wait whilst I move;" "So passes life." Sometimes short familiar proverbs were used like: "All things do wax and wane;" "The longest day must end;" "Make hay while the sun shines."

It is told of Lord Bacon, that, without intending to do so, he furnished the motto borne by a dial that stood in the old Temple Gardens in London. A young student was sent to him for a suggestion for the motto of the dial, then being built. His lordship was busy at work in his rooms when the messenger humbly and respectfully made his request. There was no answer. A second request met with equally oppressive silence and seeming ignorance of even the existence of the speaker. At last, when the petitioner ventured a third attack on the attention of the venerable chancellor, Bacon looked up and said sharply: "Sirrah, be gone about your business." "A thousand thanks, my lord," was the unexpected reply, "The very thing for the dial! Nothing could be better."

We see that the principle of the sun-dial has been recognized and utilized for many centuries; indeed, we still find sun-dials placed in gardens and parks although we rarely take the trouble to look to them for the time. Like the dinosaur and the saber-toothed tiger, they have had their day. They have been forced to give way to devices that overcame some of their objections; therefore we must not linger too long upon what is, after all, a closed chapter in the history of time-recording.

TELLING TIME BY THE WATER THIEF

Now we must take another backward step of thousands of years. In considering the subject of time-recording, it seems necessary to wear a pair of mental seven-league boots, for we must often pass back and forth over great periods at single strides. While men were still improving the sun-dial, its disadvantages were already recognized and search was being made for some other means of telling time.

Suppose, for example, that one had only a sun-dial about the house; how would one be able to tell time after sunset or on a dark day? How would one know the hour if he were surrounded by tall buildings or a thick growth of trees? And it might be very necessary to tell time under any of these conditions.

Then, again, merely as a question of accuracy, the sun-dial was not always reliable. It would get badly out of the way if used by travelers, since different markings were needed for different latitudes. While on shipboard the motion of the waves would cause the shadow to swing around in the most bewildering manner. Even under ideal conditions it was never absolutely exact, because the apparent motion of our steady-gaited old sun is not quite as dependable as most of us imagine.

Astronomers find that they must allow for what they call "equation of time" in order to make their calculations come out true. The question need not be discussed at this point, but it can be seen that, as humanity left its earliest care-free days and began to get busy, and hurried and anxious over its affairs, it came to feel that after all the sun-dial was not altogether sufficient for its needs.

For this reason we are now taking a third big backward step, returning, this time, not to the caveman but to ancient Babylon and Egypt, probably not less than twenty-seven hundred years ago and possibly much longer. In this way we meet the clepsydra.

The clepsydra was an interesting instrument, and it had an interesting name, which meant the "thief of water" and came from two Greek words meaning "thief" and "water"; you can trace this in our words "kleptomaniac" and "hydrant." We shall now examine a timepiece that was much more nearly a machine than was the simple shade-casting sun-dial.

The original idea was simple enough. At first, it was merely that of a vessel of water, having a small hole in the bottom, so that the liquid dripped out drop by drop. As the level within the jar was lowered, it showed the time upon a scale. Thus, if the hole were so small and the vessel were so large that it would require twenty-four hours for the water to drip away at an absolutely steady rate, it may be seen that the side of the vessel might easily have been marked with twenty-four divisions to indicate the hours. It may also be seen that the water would drip as rapidly at night or in shadow as in sunlight. And the clepsydra could be used indoors, which the sun-dial could not, although it required attention in that it must be regularly refilled and the orifice must always be kept completely open, because the slightest stoppage would retard the rate of dripping and the "clock" would run slow.

The sun, which, with the other heavenly bodies, had therefore been the sole reliance of the human race in its time-reckoning could now be ignored and the would-be timekeeper called to his aid another mighty servant from the forces of nature that of gravitation.

The most interesting human fact, however, about the clepsydra is that it involved an entirely different conception of the marking of time. Now it was not so much a question of when as of how long. A good sun-dial set in a proper position would always indicate three o'clock when it was three o'clock, but the clepsydra might do no such thing. It would merely show how many hours had elapsed since last it was filled, and the steady drip, drip, drip of the escaping water could and did lower the surface quite as evenly at one time of day as at another.

We have already seen that the first purpose in marking time was merely for making appointments, but the clepsydra shows that, with its invention, mankind had already made some progress toward a new point of view. One important factor in this change was the very practical need of telling time at night, in stormy weather, or indoors, where the sun-dial could not be used. The clepsydra, on the other hand, worked equally well at any hour or place, and in all sorts of weather.

Nevertheless, it, too, proved to have certain faults. After a time, people noticed the interesting fact that water ran faster from a full vessel than from one which was nearly empty; this was, of course, because of the greater pressure. Since such a variation interfered with calculations, they hit upon the idea of a double vessel; the larger one below containing a float which rose as the vessel filled, thus marking the hours upon the scale, and the smaller one above, the one from which the water dripped, being kept constantly filled to the point of overflow.

This improved form of clepsydra opened a field of fascinating possibilities in time-recording it gave the chance to make use of a machine. There is, perhaps, no more interesting point in studying human development than to see the steady, inevitable way in which mankind from its cave-dwelling days has tended toward machinery. Roughly, this progress may be characterized as of three stages.

First. Primitive man an upright-standing animal, naked, unarmed, weak as compared with some creatures, slow as compared with others, clumsy as compared with still others a creature with many physical disadvantages, but with the best brain in the animal kingdom.

Second. The tool-using man, who had begun to grasp weapons and to fashion implements, thus supplementing his natural abilities by artificial means.

Third. The machine-making man, who has fashioned to himself a mechanical "body" of incredible powers that is to say, he has learned to intensify his own powers through artificial means which he has invented, as when he made the telescope to give himself greater vision; he has made inventions by means of which he can outrun the antelopes, outfly the birds, outswim the fishes, outgaze the eagles, and overmatch the elephants in sheer physical force he can turn night into day, can send his voice across the continent, can strike crushing blows at a distance of many miles and can carry the movements of the stars in his pocket. Some phases of this third stage were foreshadowed when man first applied wheels and pulleys to his clepsydra.

Here, then, was water steadily raised or lowered by means of uniform dropping; here was a float whose motion was controlled by that of the water; here, in fact, was water-power with a means for applying it. Attach a cord to the float, cause it to turn a wheel by use of the pulley-principle, and the motion of the wheel would indicate the time. Still better, rig up a turning-pointer, increase its speed through the use of toothed gear-wheels, place it in front of a stationary disk divided to indicate the hours, and now the apparatus looked not unlike a modern clock. Or attach a bell and let it be caused to ring at a certain point in the motion what was that but an alarm-clock? Ctesibus of Alexandra was the one who is believed first to have applied the toothed wheels to the clepsydra and this was about 140 B. C.

Clepsydra

Clepsydrae were expensive of course; accurate mechanical work was never cheap until modern times. Cunning craftsmen spent their time upon costly decorations, and these water-clocks became triumphs of the jeweler's art, a gift for kings. Therefore, like the sun-dial, they drifted into Rome that vast maelstrom of the ancient world. Imagine a great walled city of low flat-roofed buildings, with fronts and porches of great columns, a town mostly of stone and much of it of marble, gleaming white under the bright Italian sun, the streets thronged with men in tunics and togas and here and there some person of importance driving by, standing erect in his chariot drawn by four horses harnessed abreast. And statues everywhere, in the streets and about the buildings and in cool courtyards and gardens among green leaves. The ancients thought of sculpture as an outdoor thing, and where we have one statue in the streets or public places of our cities, they had a hundred. We treasure the remains of them as artistic wonders in our museums, but they put them indoors and out as common ornaments, and lived among them.

Clepsydra

Presently we hear of the clepsydra being used in Roman law courts by command of Pompey, to limit the time of speakers. "This," says one writer of the day, "was to prevent babblings, that such as spoke ought to be brief in their speeches." It is not difficult to picture some pompous and tiresome togaed advocate, rolling out sonorous Latin syllables as he cites precedents and builds up arguments, while an unseen dropping checks the time against him, and to hear his indignant surprise and the chuckles of his auditors when the relentless water-clock cuts him short in the middle of some period. Martial, the Latin poet, referring to a tiresome speaker who repeatedly moistened his throat from a glass of water during the lengthy speech, suggested that it would be an equal relief to him and to his audience, if he were to drink from the clepsydra. But Roman lawyers were not guileless, and sometimes, so we are told, they tampered with the mechanical regulation or else introduced muddy water, which would run out more slowly.

This suggests one of the difficulties of the clepsydra. Still more serious was the fact that it would freeze on frosty nights. There were no Pearys among the ancient Romans; polar exploration interested them not at all; but they did spread their conquests into regions of colder weather as when Julius Caesar mentions using the clepsydra to regulate the length of the night-watches in Britain. His keen mind noted by this means that the summer nights in Britain were shorter than those at Rome, a fact now known to be due to difference of latitude.

The Clepsydra, one of the earliest time-telling devices, was used in Roman law courts to limit the time of speakers and "to prevent babbling."

As late as the ninth century, a clepsydra was regarded as a princely gift. It is said, that the good caliph, Harun-al-Raschid, beloved by all readers of the "Arabian Nights," sent one of great beauty to Charlemagne, the Emperor of the West. Its case was elaborate, and, at the stroke of each hour, small doors opened to give passage to cavaliers. After the twelfth hour these cavaliers retired into the case. The striking apparatus consisted of small balls which dropped into a resounding basin underneath.

The clepsydra appears to have been used throughout the Middle Ages in some European countries, and it lingered along in Italy and France down to the close of the fifteenth century. Some of these water-clocks were plain tin tubes; some were hollow cups, each with a tiny hole at the bottom, which were placed in water and gradually filled and sank in a definite space of time.

When the clepsydra was introduced from Egypt into Greece, and later into Rome, one was considered enough for each town and was set in the market-place or some public square. It was carefully guarded by a civic officer, who religiously filled it at stated times. The nobility of the town and the wealthy people sent their servants to find out the exact time, while the poorer inhabitants were informed occasionally by the sound of the horn which was blown by the attendant of the clepsydra to denote the hour of changing the guard. This was much in the spirit of the calls of the watchmen in old England, and later in our New England, who were, in a way, walking clocks that shouted "Eleven o'clock and all's well," or whatever might be the hour.

Allowing for the fact that the clepsydra was none too accurate at the best and that its reservoir must occasionally be refilled, it can be seen that this early form of timepiece, having played its part, was ready to step off the stage when a more practical successor should arrive.

With one of its earliest successors we are familiar.

HOW FATHER TIME GOT HIS HOUR GLASS

Every now and then one sees a picture of a lean old gentleman, with a long white beard, flowing robes, and an expression of most misleading benignity. In spite of his look of kindly good humor, he is none too popular with the human race and his methods are not always of the gentlest. In one hand he carries the familiar scythe, and, in the other, the even more familiar hour-glass. By this we may assume that he began to be pictured in this way while the hour-glass was still in common use.

The principle of the hour-glass is so similar to that of the clepsydra, and its first use was so early, that it is somewhat of a misnomer to speak of it as a successor. About the only justification that can be made is that the clepsydra has long disappeared, while the sand-glass if not the hour-glass is still sold in the stores for such familiar uses as timing the boiling of eggs, the length of telephone-conversations, and other short-time needs.

Nothing could be much simpler than the hour-glass, in which fine sand poured through a tiny hole from an upper into a lower compartment. It had none of the mechanical features of the later clepsydræ; it did not adjust itself to astronomical laws like the perfected sun-dials; it merely permitted a steady stream of fine sand to pass through an opening at a uniform rate of speed, until one of the funnel-shaped bowls had emptied itself then waited with entire unconcern until some one stood it upon its head and caused the sand to run back again.

However, it possessed some very solid advantages of its own. It would not freeze; it would not spill over; it did not need refilling; it would run at a steady rate whether the reservoir were full or nearly empty; it could be made very cheaply, and there was nothing about it to wear out.

A water-clock might be of considerable size but a sand-clock, since it required turning, must be kept small, and an hour-glass a size small enough to carry became popular, although its use was correspondingly limited. Thus, it naturally was assigned to Father Time to be carried before watches were available. A sun-dial simply would not answer this purpose, since the old gentleman works by night as steadily as by day.

How old is the sand-glass?

We do not know definitely, but it is said to have been invented at Alexandria about the middle of the third century B. C. That it was known in ancient Athens is certain, for a Greek bas-relief at the Mattei Palace in Rome, representing a marriage, shows Morpheus, the god of dreams, holding an hour-glass. The Athenians used to carry these timepieces as we do our watches.

Some hour-glasses contained mercury, but sand was an ideal substance, for, when fine and dry, it flows with an approximately constant speed whether the quantity is great or small, whereas, liquids descend more swiftly the greater the pressure above the opening.

Hour-glasses were introduced into churches in the early sixteenth century when the preachers were famous for their wearisome sermons. The story is told of one of these long-winded divines who, on a hot day, had reached his "tenthly" just as the restless congregation were gladdened to see the last grains of sand fall from the upper bowl. "Brethren," he remarked; "Let us take another glass," and he reversed it "Ahem, as I was saying " And he went on for another hour.

Other preachers, more merciful, used a half-hour glass and kept within its limits. Many churches were furnished with ornamental stands to hold the glass. These timekeepers lingered along in country churches for many years, but ceased to be in anything like general demand after about 1650.

For rough purposes of keeping time on board ship, sand-glasses were employed and it is curious to note that hour and half-hour glasses were used for this purpose in the British navy as recently as the year 1839.

The very baby of the hour-glass family was a twenty-eight second affair which assisted in determining the speed of the vessel. The log-line was divided by knots, at intervals of forty-seven feet, three inches, and this distance would go into a nautical mile as many times as twenty-eight seconds would go into an hour. When the line was thrown overboard the mariner counted the number of knots slipping through his fingers while his eyes were fixed on the tiny emptying sand-glass, and in this way so many "knots" an hour denoted the ship's speed in miles.

In the British House of Commons, even at the present time, a two-minute glass is used in the preliminary to a "division," which is a method of voting wherein the members leave their seats and go into either the affirmative or negative lobbies. While the sand is running, "division-bells" are set in motion in every part of the building to give members notice that a "division" is at hand.

It was an ancient custom to put an hour-glass, as an emblem that the sands of life had run out, into coffins at burials.

Another early means of recording time applied the principle of the consumption of some slow-burning fuel by fire. From remote ages, the Chinese and Japanese thus used ropes, knotted at regular intervals, or cylinders of glue and sawdust marked in rings, which slowly smoldered away. Alfred the Great, that noble English king of the ninth century, is said to have invented the candle-clock, because of a vow to give eight hours of the day to acts of religion, eight hours to public affairs, and eight hours to rest and recreation. He had six tapers made, each twelve inches long and divided into twelve parts, or inches, colored alternately black and white. Three of these parts were burned in one hour, making each inch represent twenty minutes, so that his six candles, lighted one after the other by his chaplains, would burn for twenty-four hours.

The Eskimos also, through the long arctic night have watched the lamp which gives both light and heat to their cold huts of snow. But all these are no more than crude conveniences, whose irregularity is evident, and there is likewise no need to do more than call attention to the effect upon fire in any form, of wind or dampness in the air. The Roman lamp-clock sheltered from the weather was the best of them all, and was the only one which long continued in civilized use.

Our chief interest in all such devices comes from the touch of poetry still remaining in the tradition of the sacred flame which must be kept forever burning, and in association of life and time with fire, in such parables as that of the Wise and Foolish Virgins. There is a reminder of this old time-keeping by fire in all that poetry and philosophy which tells of hope that still may live or of deeds that maybe done, "while the lamp holds out to burn."

Thus far, in spite of occasional glimpses of the Middle Ages and of modern times, we have dealt, for the most part, with earlier ages. Now our story must leave these behind, and thus passes the ancient world with its strange pagan civilization which was so human, so wise and so simple. It is difficult for modern Americans even to imagine existence in ancient Greece or Rome or in still more ancient Egypt and Mesopotamia since the whole attitude toward life was so essentially different from what it is to-day.

Our debt to the ancients in this one matter of recording time is typical of that in many others. To them we owe our whole fundamental system and conception of it from the astronomy by which we measure our years and our seasons and make our appeal to the final standard of the stars, down to the arithmetic of our minutes and seconds and the very names of our months and days.

The sun dial is the first ancestor of all time tellers, and the sand glass was probably the first portable time telling device.

In the modern application and practical use of all this, on the other hand, we owe them nothing. They never made a clock or watch, or any like device which has more than a merely ornamental use to-day. They gave us the general plan so well that we have never bettered it, but they left later generations to work out the details. They invented the second as a division of time but they did not measure by it. They did not care to try. For them, learning was the natural right and power of the few, and the gulf between the most that was known by the few and the little that was known in general, was like the gulf between great wealth and great poverty among ourselves.

Indeed, in this age of teaching and preaching, when a thought seems to need only to be born in order to be spread abroad over the world, it is hard for us even to conceive the instinct by which men kept their learning like a secret among the initiated and felt no impulse to make known that which they knew.

Their great men thought and did wonderful things which are now the common property of us all. And their common folk lived in a fashion astonishingly primitive by comparison, in an ignorance which certainly was weakness and may somehow have been bliss.

That world of theirs is gone the body and the spirit of it alike. And there remains to us, along with much of their art and their science, the hour-glass to symbolize that relentless flight of time which they feared but never tried to save; and the quaint sun-dial in our gardens, a memory of that worldly-wise old philosophy which counted only the shining hours.

THE CLOCKS WHICH NAMED THEMSELVES

Now the scene changes again, and the story shifts forward over the interval of a thousand years. As we take up the tale once more, we find ourselves in another world, amid a life as different from that ancient life of which we have been speaking as either of them is from our own life to-day.

The ancient civilization, which may be traced from Rome through Greece, Babylon and Egypt back to the dim dawn of history, is gone almost as if it had never been. For there came a period when great hordes of barbarians defeated the armies, burnt the cities, pillaged and destroyed, leaving only desolation and ruin behind them. Then followed hundreds of years of what we call the "Dark Ages," ages of ignorance and violence, when mankind was slowly struggling upwards again and was forming a new civilization upon the ruins of the old. Therefore, at the point we have now reached, there are no more white temples and pillared porticos and sandaled men in white tunic and toga, and marble statues in green gardens; but everywhere we find sharp roofs and towers, quaint outlines, and wild color like a child's picture-book.

There are castles with their moats and battlements, and monasteries with their cloistered arches; there are knights in armor riding, and lords and ladies gorgeous in strange garments, and monks in their dull gowns, and the sturdy peasant working in the field; and in the towns, all among peaked gables and Gothic windows and rough cobbled streets, a motley crowd of beggar and burgher and courtier, priest and clerk, doctor and scholar and soldier and merchant and tradesman an endless variety of types, and each in the distinctive costume of his calling. And there are churches everywhere, from the huge cathedral towering like a forest of carven stone to the humble village chapel or wayside shrine, their spires all pointing up to heaven in token of the change that has come upon the life and spirit of the world.

We have come from the height of the classic period suddenly into the heart of the Middle Ages; and in the dark centuries that lie between, Christ and His Disciples have come and gone, and the religion of the Western World has changed; the old gods have perished and the saints have filled their places. And Rome has died, and Romance has been born.

The center of civilization has shifted to the north and west; from the old ring of lands around the Mediterranean to the great nations of modern Europe. Italy has become a jealous group of independent cities, great in art and commerce, but in little else. Germany is much the same, except for the lack of some few score centuries of tradition. France and Spain are already great and growing. William the Conqueror has fought and ruled and died, and the "Merry England" of song and story has grown up out of the fusion of Saxon and Norman. Chivalry and the Crusades, the times of Ivanhoe and The Talisman, are as fresh as yesterday.

And by green hedgerows and hospitable inns, Chaucer's Pilgrims are plodding onward toward the sound of Canterbury's bells. For here is the point of all our seeking that there are clocks now in the monasteries and in the Cathedral towers. There is just one curious link of likeness between the Middle Ages and the remoter past; as it was at first at Babylon, so now in the fourteenth century the priesthood holds almost a monopoly of science and of learning.

Thus, although the sun-dial, clepsydra and sand-glass are still much used, we find ourselves at last in the time and lands of clocks. The very sound of the word "clock" gives a clue to its origin. It suggests the striking of the hour upon some bell. The French called the word cloche and the Saxons clugga, and both of these originally meant a bell.

If you will put yourself back in the picture at the beginning of the chapter, you will find yourself in a realm of sounding, pealing, chiming bells with the hours of prayer throughout the day, from matins to angelus, rung out from the belfries, and with frequent deep-toned strikings of the hour. Not even a blind man could have remained unconscious of the passage of the hours under such conditions, and time, in a sense, became more a possession of democracy although timepieces themselves were still the mark of special privilege.

Life also was beginning to hurry just a little. Very deliberate, we should call it in comparison with the mad rush of the twentieth century, and yet it began to show its growing complexity in that humanity was becoming more definitely organized and men were forced to depend more and more upon each other. In all of this, there was a slightly growing sense of the things that were to be, just as the water for some miles above Niagara begins to hasten its course under the influence of the mighty cataract over which it will at last go madly plunging.

Herein occurs another of those baffling questions, like the old-time puzzler as to whether the hen first came from the egg or the egg from the hen. One cannot help wondering to what extent the increasing accuracy of the broadening knowledge of time-keeping was the result of our complicated modern life and to what extent it was the cause. Certainly we cannot conceive of present-day affairs as being conducted save in the light of moving hands and figures upon a dial.

From the Middle Ages, then, we get our word for clock and, which is more important, we begin to get some crude application of its modern mechanical principles. They were wonderfully skilful, those medieval workmen, considering the means at their disposal, and the ingenuity of some of their clocks is still a delight, but, perhaps, for better understanding of the story, we should stop for a minute to inquire exactly what a clock means from the mechanical point of view.

A clock is a machine for keeping time. And for this there are four essentials, without any one of which there would be no clock. First, there must be a motive power to make it run; second, there must be a means of transmitting this power; third, there must be a regulating device to make the mechanism move steadily and slowly, and keep the motive power from running down too quickly; and, fourth, there must be some device to mark the time and make it known.

In a typical modern clock the power comes from the pull of a weight or the pressure of a spring although clocks may, of course, be operated by electricity or compressed air or some other means; also, the regulator is what is known as the "escapement" and the recording device consists of the hands, the dial, and the striking mechanism. Having stated this, let us return to the past and see if we can determine how these principles came to be applied.

This is not altogether easy. Our forefathers were less particular than we over such trifling questions as names and spelling even the learned Shakespeare, long afterward, used several different spellings of his own name. Thus, when we see in the records of the period the name of "clock" or "horologe" we cannot tell with certainty what type is meant, since "horologe" meant simply a device for keeping time; it might have been applied equally well to a clock, clepsydra, an hour-glass, or even a sun-dial.

"It is quite possible," writes M. Gubelin Breitschmidt, the younger, an eminent horologist of Lucerne, Switzerland, "that a large number of the technical inventions of antiquity were lost during the migrations of the barbarians and under the chaotic conditions prevailing during the first thousand years of Christianity, but the most perfect surviving instrument for measuring time was the water-clock, known as the clepsydra, which was able to maintain its supremacy long after the appearance of the wholly mechanical clock, just as the beautiful manuscripts of the artist monks and laymen were favored by the cultured classes long after the invention of movable types for printing.

"The spread of Christianity throughout Europe caused the foundation of many religious communities, and the severe rules by which they were governed fixing the hours of prayer, labor, and refreshment forced their members to seek instruments by which to measure time. In the year 605, a bull of Pope Sabinianus decreed that all bells be rung seven times in the twenty-four hours, at fixed moments and regularly, and these fixed times became known as the seven canonical hours. The sound of the bells penetrated and came to regulate not only the life of the religious bodies but also that of the secular people who lived outside the walls of the monasteries. Oil-lamps, candles, hour-glasses, prayers and for those who had the means of buying them clepsydræ served as chronometers for the brotherhoods; so that one can easily imagine that many a monk sought to improve these instruments. But as yet, no one had found means to regulate the wheel-system of a movement. In the best instruments of this period, water supplied the motive power and served as well to regulate the action."

There is a general belief that Gerbert, the monk, who was the most accomplished scholar of his age, and who later became Pope Sylvester II, was the one who first took the important step of producing a real clock, and that this occurred near the close of the tenth century or to be more exact, about 990 A. D. This period was one of densest superstition, and expectancy of the end of the world was in the air, since many people had fixed upon the year 1000 A. D. as the date of that cataclysmic event.

Authorities of the Church and of the state were not very partial to invention and research, their attention being fixed largely upon theological, political, or military affairs; but, of course, inquiring and constructive minds were still to be found; even without encouragement these tended to follow the impulse of their natures.

As a youth of seventeen Galileo watched a swinging lamp, in the Cathedral of Pisa, timed it by his pulse, and discovered the principle upon which pendulum clocks are built.

It is to the monks in their cloisters that we chiefly owe the preservation of learning through the "dark ages," and from the monks, for the most part, came such progress of science and invention as was made. If Gerbert, the monk, after patient tinkering with wheels and weights in his stone-walled workshop, really achieved some form of the clock-action as we know it, he was one of the great benefactors of the human race. Still, it is not impossible that his device may only have been a more remarkable application of the clepsydra principle.

Whatever it was, it seems to have startled the authorities, for they are said to have accused him of having practiced sorcery through league with the devil, and to have banished him for a time from France. His age appears to have had a vast respect for the intellectual powers of his Satanic Majesty. Anything which was too ingenious or scientific to be understood without an uncomfortable degree of mental application was very apt to be ascribed to diabolic inspiration and thus found unfit for use in "Christian" lands. It could hardly have been a stimulating atmosphere for would-be inventors.

All of the credit that we are ascribing to Gerbert must therefore be prefixed with an "if." Did he really invent the clock-movements, or is this merely another of the tales which have blown down to us from this age of tradition and romance? For similar tales are told of Pacificus in 849 A. D. of the early Pope Sabinianus in 612 and even of Boetheus, the philosopher, as far back as 510 A. D., while always in the background are claims of priority for the Chinese who are supposed to have discovered many of our most important mechanical and scientific principles away off upon the other side of the world before these were dreamed of in the west.

If all of these various claims were true, which is far from likely, it still would not need to surprise us, for it must be remembered that humanity, until within the past few generations, was more or less a collection of separated units and its records were very incomplete. There was scant interest in abstract research and very limited intercourse between towns and countries; one who made an important discovery in one locality might be unheard of a hundred miles away. Unless all the conditions were favorable, his ideas might even pass from memory with his death, until some scholar of modern times might chance upon their record.

All that can with certainty be said, therefore, is that there were clocks of some sort in the monasteries during the eleventh century; that back of these were the clepsydræ and other time recording devices; and that here and there through the preceding centuries are more or less believable tales of inventions that had to do with the subject.

Let it be remembered, too, that some of the brilliant minds of ancient times made discoveries that were forgotten after the barbarian waves overwhelmed preceding civilizations. The ages following the downfall of Rome were those of intellectual darkness, illiteracy, and rude force until mankind groped slowly back toward the light through the process of rediscovery.

Thus, it mattered not at all to the medieval world that Archimedes, the great Greek scientist and engineer who, however, chanced to live in the Greek colony of Sicily was able, somewhere about 200 B. C., to construct a system of revolving spheres which reproduced the motion of the heavenly bodies. Such a machine must necessarily have involved some sort of clock-work. We dare not stop to consider Archimedes, lest we stray too far from our subject, but this marvelous man of ancient times, the Benjamin Franklin of his day, seems to have had a hand in almost every sort of mechanical and scientific research, from discovering the principle of specific gravity, in order to checkmate a dishonest goldsmith, to destroying Roman war-ships by means of his scientific "engines." The story is told that he set the ships on fire by concentrating upon them the rays of the sun from a number of concave mirrors. And, although this story may not be true, the things that he is known to have done are extraordinary.

Archimedes and his knowledge had long passed away when the monastery clocks of the eleventh century began to sound the hour. These were the fruit of a crude new civilization just struggling for expression, and represented the general period when William the Conqueror led his Norman army into England.

THE MODERN CLOCK AND ITS CREATORS

We learn that toward the close of the thirteenth century a clock was set up in St. Paul's Cathedral in London (1286); one in Westminster, by 1288; and one in Canterbury Cathedral, by 1292. The Westminster clock and the chime of bells were put up from funds raised by a fine imposed on a chief justice who had offended the government. The clock bore as an inscription the words of Virgil: "Discite justitiam moniti," "Learn justice from my advice," and the bells were gambled away by Henry VIII! In the same century, Dante, whose wonderful poem the Commedia, (the Inferno, Purgatory and Paradise) is sometimes called the "Swan Song of the Middle Ages," since it marks the passing of the medieval times, spoke of "wheels that wound their circle in an orloge."

Chaucer speaks of a cock crowing as regularly "as a clock in an abbey orloge." And this shows, curiously, the early meaning of the word, for by the word "clock," Chaucer evidently meant the bell which struck the hour, and, very obviously, he used the word "orloge" to indicate the clock itself.

Many of these "clocks" had neither dials nor hands. They told time only by striking the hour. Sometimes in the great tower clocks there were placed automatic figures representing men in armor or even mere grotesque figures which, at the right moment, beat upon the bell. These figures were called "jacks o' the clock" or "jacquemarts" and curious specimens of them are still in existence.

The early abbey clocks did not even strike the hour but rang an alarm to awaken the monks for prayers. Here again, the alarm principle precedes the visible measurement of time; even now, as already noted, we speak of a "clock" by the old word for "bell."

In the course of the following century the fourteenth clocks began to appear which were really worthy of the name, and of these we have authentic details. They were to be found in many lands. One of them was built, in 1344, by Giacomo Dondi at Padua, Italy. Another was constructed in England, in 1340, by Peter Lightfoot, a monk of Glastonbury. And in 1364, Henry de Wieck, De Wyck, or de Vick, of Wurtemburg, was sent for by Charles V, King of France, to come to Paris and build a clock for the tower of the royal palace, which is now the Palais de Justice. It was finished and set up in February 1379, and there it still remains after lapse of five and a half centuries, although its present architectural surroundings were not finished until a much later date.

This venerable timepiece termed by some chroniclers "the parent of modern timekeepers," was still performing its duty as late as 1850. And so it is a matter of interesting record that its mechanism, which served to measure the passage of time in the days when the earth was generally believed to be flat and when the Eastern Division of the Roman Empire was still ruled from Byzantium, now Constantinople, has served the same purpose within the possible memory of men now living. Its bell has one grim association it gave the signal for that frightful piece of Medicean treachery, the Massacre of St. Bartholomew, planned by Catherine de Medici, the mother of the King Charles IX, when the armed retainers of the crown of France flung themselves upon the unsuspecting Huguenots and caused the streets to run red with the blood of men, women and children a ghastly butchery of thousands of people.

As we have seen, de Vick's clock was neither the earliest made, nor among the earliest; nor, probably, did it embody any at that time new mechanical invention. It does, however, fairly and clearly typify the oldest style of clock of which we to-day have any accurate knowledge. Compare its description, then, with the clock upon your shelf.

We think of the tall-cased "grandfather's clocks" as antique; but this tower-clock of de Vick's outdoes them in antiquity by some four hundred years. And its most interesting feature is its curious likeness in mechanical principle to the clocks of modern times. Like most early clocks, it has only one hand the hour-hand. Its ponderous movement is of iron, laboriously hand-wrought; the teeth of its wheels and pinions were cut out one by one. It was driven by a weight of five hundred pounds, the cord of which was wound round a drum, or barrel. This barrel carried, at one end, a pinion, meshing with the hour-wheel, which drove the hands; the flange at the other end of the barrel formed the great wheel, or first wheel of the train. This meshed with a pinion on the shaft of the second wheel, and this in turn with a lantern-pinion upon the shaft of the escape-wheel. All of this is, of course, essentially the modern train of gears, only with fewer wheels.

The escapement is the most important part of the whole mechanism, because it is the part which makes the clock keep time. It is an interrupter, checking the movement almost as soon as, under the urge of the mainspring, it starts forward. The frequency and duration of these interruptions determines the rate of running. Without this, the movement would run down swiftly; with it, the operation stretches over thirty hours, involving 432,000 interruptions.

The huge and elaborate Clock of Strasbourg Cathedral, in Lorraine, was built in 1352 and is an example of the first clocks.

De Vick's escapement is shown in the illustration. The escape-wheel was bent into the shape of a shallow pan, so that its toothed edge was at a right angle to the flat part of the wheel. Near it was placed a verge, or rotating shaft, so called from a Latin word meaning "turning around." On this verge were fastened two flat projections called pallets, diverging from each other at about an angle of one hundred degrees. The width between the pallets, from center to center of each, was equal to the diameter of the wheel, so that one would mesh with the teeth at the top of the escape-wheel and the other with the teeth at the bottom.

de Vick's Clock
de Vick's Clock

Now, if the upper pallet were between the teeth at the top of the wheel, the pressure of the wheel trying to turn would push it away until the teeth were set free. But, in so doing, it would cause the verge to turn and bring the lower pallet between the teeth at the bottom of the wheel. And since the bottom of the wheel was, of course, traveling in the opposite direction from the top, the action would be reversed, and the lower pallet would be pushed away, bringing the upper one back between the teeth of the wheel again; and so on, "tick-tock," the wheel moving a little way each time, and the pallets alternately catching and holding it from going too far.

The device was kept running slowly by means of a cross-bar called a "foliot," fastened across the top of the verge in the shape of a T, and having weights on its two ends. When this weighted bar was set turning in one direction, it would, of course, resist being suddenly stopped and started turning the other way, as it was constantly made to do. And this furnished the regulating action which retarded the motion of the works and kept them from running down.

This involves the principle of the modern balance-wheel in both watches and clocks, which is that of inertia; the rim of the balance-wheel represents the weights on the bar that resist the pull of the pallets. A vital improvement, however, is the interception of the hair spring which gives elasticity to the pull and thus supplies the elements of precision and refinement. The inertia of the balance-wheel is gauged by the weight of the rim and its distance from the center; and the last refinement of regulation of the mechanism is produced by moving the tiny screws on the periphery of this wheel outward or inward.

We shall see later how this old escapement was in principle much like the improved forms in use to-day. It was as quaint and clumsy an affair as the first automobile or the first steam-engine. But, like them, it was a great invention, destined to achieve great results. For it was the means of making a machine keep time. And every clock and watch in use to-day depends for its usefulness upon a similar device. The tick is the first thing we think of in connection with a clock; and it is the most essential thing also, because it is the escapement which does the ticking.

This old clock of de Vick's also struck the hours upon a bell and in very much the same way as modern clocks are made to do. But the mechanical means by which it did so are too complicated to be easily described here. And indeed it is unnecessary to do so, since the bell is far less important. A clock need not strike, but it must keep time.

On the fearsome eve of St. Bartholomew, therefore, and again within the past generation, the clanging of this old clock's bell was brought about by the whirling gears and ponderous weights of an early craftsman who wrought his work into the ages.

As already stated, de Vick's mechanism embodied mechanical principles which, although greatly developed and improved, are employed even at the present day. All the essentials of a clock are there; the motive power the descent of a massive weight is now replaced by a slender spring; the train of gears by which this motion is reduced and communicated, are cut to-day with the extreme accuracy of modern machine work; the hand moving around the dial is now accompanied by a longer, swifter hand to tell the minutes; the escapement which by checking the motive power while yet allowing it to move on step by step, retards and regulates even the numbered striking of the unchanging hours.

De Vick's old clock may have been a crude machine it certainly was a poor timekeeper but it was the sturdy ancestor of all those myriad tribes of clocks and watches which warn us solemnly from our towers, chime to us from our mantels, or, nestling snugly in our pockets, or clinging to our wrists, help us to maintain our efficiency in the complexities of modern life. The mechanism employed by de Vick was retained without any improvement of importance in all the time-pieces of the next three hundred years. The foliot escapement, especially, remained in use much longer. Indeed, any modern watchmaker would recognize that it was practically a horizontal balance-wheel.

Long before it was improved upon, watches had been invented and clocks had everywhere become common. But we shall reserve the watch for the next chapter; for the moment, our concern is with clocks alone.

The disadvantage of the medieval clock was its inaccuracy. This was due first to crude workmanship and unnecessary friction; but that trouble was presently overcome, for the medieval mechanic could be as fine and accurate a workman as any modern. He had the artist's personal pride and pleasure in his skill, and also a great unhurried patience, somewhat hard for us to picture in this breathless age. At best, however, his work fell far short of the accuracy possible with modern machinery. Other important difficulties were found in the expansion and contraction of parts due to temperature variations, and the fact that the foliot balance was at its best only when running slowly. Altogether, then, these early clocks were easily surpassed in accuracy of timekeeping by a sun-dial or a good clepsydra.

The question arises, therefore, why this newcomer in the field of timekeeping, should have begun to displace the earlier devices. The clock was not yet a better timepiece than the sun-dial; why did it grow more common? Well, for one thing, people like novelties. For another, people loved their churches and lived by the chimes of distant bells; and the clock was by far the most practical striking device, whatever might be its faults in keeping time. But, what was most important of all, it was a machine, susceptible of infinite improvement and offering a field for endless ingenuity. It appealed to that inborn mechanical instinct by means of which mankind has wrought his mastery over the world.

We have seen how de Vick's clock contained, as it were, the germ of all our clocks. And, moreover, the medieval regarded machinery with profoundest awe. It is the unknown which awakes imagination. We wonder at the cathedrals of his day, but the medieval knew about cathedrals; he built them. Considering their comparatively cruder tools, lack of modern hoisting machinery, and so forth, their architectural and building abilities exceeded even those of to-day. On the other hand, a locomotive or a modern watch, such as we glance at without special notice, would have appeared to him the product of sheer sorcery, too wonderful to be the work of human hands.

The Middle Ages could not much improve their clock without some radical invention; and such a mechanical type of invention was yet the province of but few minds. The typical craftsman could merely make the clock more convenient, more decorative, and more wonderful. To this work, he and his fellows addressed themselves with all of their patient skill and their endless ingenuity for ornamentation.

They made clocks for their churches and public buildings, and elaborated them with intricate mechanical devices. The old "Jacks" that struck the bells were only a beginning. They made clocks for their kings and wealthy nobles, adorning them with all the richness that an artist could design and a skilful jeweler execute. They made clocks even for ordinary domestic use so quaint in design and so clever in workmanship that we exhibit them to-day in our museums. One difficulty in determining the date of the first invention is that long before the days of de Vick and Lightfoot, machines were made to show the day of the week and month and to imitate the movements of the stars; and the first horological records may refer to clock-works of this kind.

The famous clock of Strassburg Cathedral shows the extreme to which the medieval craftsman carried this kind of ingenuity. It was originally put up in 1352 and has been twice rebuilt, each time with greater elaboration. It is three stories high and stands against the wall somewhat in the shape of a great altar with three towers. Among its movements are a celestial globe showing the positions of the sun, moon, and stars, a perpetual calendar, a device for predicting eclipses and a procession of figures representing the pagan gods from whom the days of the week are named. There are devices for showing the age and phases of the moon and other astronomical events. The hours are struck by a succession of automatic figures, and at the stroke of noon a cock, perched upon one of the towers, flaps his wings, ruffles his neck, and crows three times. This clock still remains, having last been rebuilt in the four years 1838 to 1842. But its chief interest is that of a mechanical curiosity. It keeps no better time than a common alarm-clock, nor ever did. And in beauty as well as usefulness, it has been surpassed many times by later and simpler structures.

For the first really important improvement in clock making we must pass to the latter end of the sixteenth century. The Italian Renaissance with its great impulse to art and science has come and gone, and the march of events has brought us well into the modern world. America had been discovered a century and is beginning to be colonized. Spain is trying to found a world empire upon blood and gold and the tortures of the Inquisition. England is at the height of the great Elizabethan period. It is the time of Drake and Shakespeare and Sir Walter Raleigh.

At this period of intellectual awakening, a remarkable young man steps upon the scene. In 1564, the year in which the wonderful Englishman, Shakespeare, first saw the light of day, the scarcely less wonderful Italian, Galileo, was born in Pisa. He was gifted with keen eyes and a swift, logical mind, which left its impress upon so many subjects of human thought and speculation that we are tempted to stop as with Archimedes and trace his history. But, one single incident must suffice.

In 1581, this youth of seventeen stood in the cathedral of Pisa. Close at hand, a lamp suspended by a long chain swung lazily in the air currents. There was nothing unusual in such a sight. Millions of other eyes had seen other suspended objects going through exactly this motion and had not given the sight a second thought. At this moment, however, a great discovery of far-reaching application one which was to revolutionize clock construction hung waiting in the air. Young Galileo took notice.

The lamp swung to and fro, to and fro. Sometimes it moved but slightly. Again, as a stronger breeze blew through the great drafty structure, it swung in a considerable arc, but always and this was the point which impressed itself upon the Italian lad the swing was accomplished in exactly the same time. When it moved a short distance, it moved slowly; the farther it moved, the faster became the motion; in its arc it moved more swiftly, accomplishing the long swing in the same time as it did the short one. In order to make sure of this fact, Galileo is said to have timed the swinging lamp by counting the beating of his pulse.

Thus was discovered the principle of the pendulum and its "isochronism." By "isochronism" we mean inequal arcs in equal time. In other words, any swinging body, such as a pendulum, is said to be "isochronous" when it describes long or short arcs in equal lengths of time. This also applies to a balance-wheel, and hair-spring. And herein lies a remarkable fact this epoch-making discovery was after all but a rediscovery. The isochronism of a swinging body was known in Babylon thousands of years before, although the Babylonians, of course, could not explain it. Lacking in application, it had passed from the minds of men, and it remained for Galileo to observe the long-forgotten fact and to work out its mechanical application. He did not himself apply this principle to clock-making, although some fifty years later, toward the end of his life, he did suggest such an application.

The first pendulum clocks were probably made about 1665, by Christian Huyghens, the celebrated Dutch astronomer and mathematician who discovered the rings of Saturn; and by the English inventor, Doctor Robert Hooke. The invention is claimed for several other men in England and abroad at about the same time; but hardly upon sufficient authority.

From that time on, the important improvements of clockwork were chiefly made in two directions those of the mechanical perfection of the escapement and the compensation for changes of temperature.

There is a little world of invention and discovery behind the face of the clock which beats so steadily on your mantel. Look within if you will, and see the compact mechanism with its toothed gears, its coiled spring, or its swinging pendulum, in which the motion of the cathedral lamp is harnessed for your service, nothing in that grouping has merely happened so. You may or may not understand all the action of its parts, or the technical names of them; but each feature in the structure has been the result of study and experiment, as when Huyghens hung the pendulum from a separate point and connected it with a forked crank astride the pendulum shaft. You can see that forked crank to this day, if you care to look; it was the product of good Dutch brains.

Next we come to one of the greatest single improvements in clock-work, and the chief difference between the mechanism made by de Vick and the better ones of our own time. When the pallets in a clock are forced by an increased swing of the pendulum or by the form of the pallet faces against the teeth of the escape-wheel in the direction opposite to that in which the wheel is moving, the wheel must be pushed backward a little way each time, and the whole clock action is made to back up a little. You can see that this would tend to interfere with good and regular timekeeping. George

Graham, in London, in 1690 corrected this error by inventing the dead-beat escapement which rather contradicted its name by working very well and faithfully.

There are many forms of this escapement and there is no need to explain it in detail. But the main idea is this: At the end of each vibration or swing of the pendulum, the escape-teeth, instead of being made to recoil by the downward motion of the pallets, simply remains stationary or at rest until the commencement of the return swing of the pendulum. This was brought about by applying certain curves to the acting faces of the pallets. But the acting faces of both tooth and pallet are beveled, so that the tooth in slipping by gives the pallet a "kick" or impulse outward and keeps it in motion. Nowadays, even a common alarm-clock has an escapement working in this way.

Then came another remarkably interesting contribution. Have you ever wondered why the pendulums of fine clocks were weighted with a gridiron of alternate rods of brass and steel? For purpose of ornament? Not at all it constitutes a scientific solution of an embarrasing problem, due to the inevitable variations in temperature. Metals expand with heat and contract with cold. Notched iron bars can be made to "crawl" along a flat surface by alternately heating and cooling them. Bridge-builders sometimes arrange sliding points, or rocking points to adjust the differences in the length of the steel. Contraction and expansion are important factors in all their calculations. But a pendulum would change its rate of motion if it changed its length and this would interfere with its accuracy as a measurer of time. Graham worked upon this problem, too, and attached a jar of mercury to the rod of his pendulum for a weight. When the heat lengthened the rod, it also caused the mercury to rise, just as in a thermometer, and this left the "working-length" the same.

Such mercury-weighted pendulums are not uncommon to this day, but the more familiar gridiron came from the brain of John Harrison, who, in 1726, fixed the alternate rods in such a way that the expanding brass rods raised the weight as much as the expanding steel rods lowered it. Thus they neutralized each other.

The clock as we know it was now virtually complete. There were structural refinements, but no more radical improvements to be made. In tracing its development from the fourteenth to the eighteenth century, we note one curious likeness to the ancient history of recorded time. In this case, as before in Babylon, the people first concerned with the science were the priests, and after them the astronomers, but we note a still more important difference.

As the medieval passed into the modern, the practise of horology passed more and more out of the hands of scientists into the keeping of commercial workmen. The custodian of time was at first a priest, and finally a manufacturer. And this change was attended by a vast increase in the general use of timepieces, and the correspondingly greater influence of time upon society and men's way of living. The Middle Ages made clocks and watches; and clocks and watches make the age in which we live.

THE WATCH THAT WAS HATCHED FROM THE NUREMBURG EGG

In the second act of Shakespeare's play, As You Like It, when Touchstone, the fool, meets Jaques, the sage, he draws forth a sun-dial from his pocket and begins to moralize upon Time.

Touchstone's dial must have looked like a napkin-ring, with a stem like that of a watch, by which to hold it up edgewise toward the sun, and a tiny hole in the upper part of the ring through which a little sunbeam could fall upon the inner surface whereon the hours were marked. This pinhole was perhaps pierced through a slide, which could be adjusted up or down according to the sun's position at the time of year. In principle, therefore, it was a miniature of the huge dial of Ahaz of more than two thousand years before.

In another Shakespeare play, Twelfth Night, Malvolio is gloating in imagination over his coming luxury when he shall have married the heiress and entered upon a life of wealth and leisure.

"I frown the while," says he; "and perchance wind up my watch, or play with my some rich jewel."

There, in those two quotations, we have the whole meaning of the watch in the time of Queen Elizabeth. Touchstone's dial was a practical convenience a thing to tell the time. Malvolio's watch was a piece of jewelry, an ornament indicating wealth and splendor. While watches had been well known for many years, people wore them chiefly for display and told time by means of pocket sun-dials.

For the first watches we must go back to about the year 1500, shortly after America had been discovered, and when the great tower-clocks of de Vick and Lightfoot were not much more than a century old. In the quaint old town of Nuremberg there lived, at that time, one Peter Henlein, probably a locksmith. But a locksmith, in those days, would be an expert mechanic more like a modern toolmaker; very likely an armorer also; capable of that fine workmanship in metal which we still wonder at in our museums. Nuremberg was then very much a medieval city, all red-tiled roofs and queer windows, where people went about dressed in trunks and jerkins and pointed caps and pointed shoes. It looked like Die Meistersinger, and Grimm's Fairy Tales, and pictures by Howard Pyle and Maxfield Parrish; very much like "Spotless Town," except that it was far from spotless.

Now, as you remember, there was not until long after this any means of making clocks keep anything like accurate time; so, instead of improving them, people competed with each other in devising novel and ingenious forms. There could be no more desirable novelty than a clock small enough to stand upon a desk or table, or even to be carried around. Such a clock could not well be driven by weights. But Peter Henlein overcame that difficulty by using for the motive power a coiled mainspring wound up with a ratchet, just as we still do to-day.

There is some dispute over attributing to Henlein the credit for this invention; but at least he did the thing, and it cannot be proved that anybody did it before him. "Every day," wrote Johannes Coeuleus, in 1511, "produces more ingenious inventions. A clever and comparatively young man Peter Henlein creates works that are the admiration of leading mathematicians, for, out of a little iron he constructs clocks with numerous wheels, which, without any impulse and in any position, indicate time for forty hours and strike, and which can be carried in the purse as well as in the pocket."

In Shakespear's play, "As You Like It," Touchstone, the Fool, draws forth a pocket sun dial, which probably was of the "napkin ring" type.

There was, however, no invention of any such thing as we mean by the term watch to-day that came complete from the mind of any one man, but the contrivance gradually grew, in shape and structure out of the small clock which could be worn at the belt or on a chain round the neck. It came to be called a watch because clock meant a bell that struck the hours. But many of the first watches had striking apparatus, and this circumstance added to the confusion of names. We slangily call a fat, old-fashioned watch a turnip; but the first watches were very much fatter and more old-fashioned, and might fairly have deserved the name. Before long, Henlein was making them oval in shape. Hence, they were called Nuremberg eggs.

Here, then, is something which we can really consider a watch. Let us see how it compares with those that we know to-day. In the first place, being egg-shaped, it was thick and heavy you would not like to carry it in your pocket. It had no crystal and only one hand the hour-hand. So much for the outside.

Inside, the difference was still greater. The works were made of iron and put together with pins and rivets. It was all hand-work expert workmanship, indeed but look at the works of your own watch and try to imagine cutting the teeth in those tiny gears, or making those delicate springs with files and hammers. As pieces of hand-workmanship, therefore, the watches made by Henlein and his followers were remarkable; but when compared with our modern watches, they were crude and clumsy affairs.

Furthermore, they were poor timekeepers. They had the old foliot balance running parallel to the dial. This was all very well as long as the watch lay on the table with the balance swinging horizontally. But as soon as it was carried, in a perpendicular position, the arms of the balance had to swing up and down, which was quite another matter. And then, of course, the crudeness of the works produced a great deal of friction. This made it necessary to use a very stiff mainspring, otherwise the watch would not run at all. Such a spring exercised more pressure when fully wound than when it was nearly run down. And so the worst fault of the foliot was that it speeded up under increased pressure.

The first improvements, and, in fact, the only ones for nearly two hundred years, were directed toward doing away with the unequal pressure of the mainspring and thus make the watch keep better time. If you look into the back of a very early watch, you may see a curious device consisting of a curved arm ending in a pinion, which travels round an eccentric gear of peculiar shape. This is the first type of equalizing mechanism; it was invented in Peter Henlein's time and was called the stackfreed; but it was a clumsy device at best and a great waste of power. Therefore it was gradually displaced by the fusee.

Perhaps one might have felt a certain amount of pride in carrying about such a thick, bulging mechanical toy, as were these early watches, but, as to possessing something that would keep correct time that was a different matter. After admiring it and listening to its ticking, one would have to guess as to just how far wrong it might be. People did not figure closely on minutes and half minutes in the day of the Nuremberg egg; there was no "Wall Street" and no commuting. And this brings us to a real event in the whole story.

Jacob Zech, a Swiss mechanic, living at Prague in Bohemia, Austria, about 1525, began studying the problem of the equalization of watch mechanism. He was sure that there ought to be some better means than that of the clumsy stackfreed. Presently he hit upon the principle of the fusee, and Gruet, another Swiss, perfected it. At last it became possible to make a watch that would not run fast when first wound and then go more and more slowly as it ran down and to do this in a really practical way. Before this time, a watch was a clumsy piece of ticking jewelry; now it became something of a real time-keeper. Therefore, it was not long before people began to want Swiss watches. These were the days when skilful Swiss craftsmen worked patiently in their little home shops, making some single watch-part and making it extremely well, while the so-called "manufacturer" bought up these separate parts, and assembled them into watches.

What was the fusee that brought about such a change? Not much to look at, surely merely a short cone with a spiral groove running about it, and a cord, or chain, wound in this groove and fastened at the large end of the core. Its principle and its action were very simple, and that is why it was a great invention. Some one has said that anyone can invent a complicated machine to do a piece of work, but it takes real brains to make a simple machine that will do the same work.

Stackfreed, Mainspring Barrel and Fusee

The shaft of the fusee was attached to the great wheel which drove the gears, and the other end of the cord was fastened to the mainspring barrel. This is the way in which it worked: The mainspring slowly turned the barrel; this gradually unwound the cord from the fusee and caused the fusee to turn. When the fusee turned, the wheels also were forced to turn, and the watch was running. At the start, the cord would unwind from the small end where the leverage was least, but as the tension of the mainspring grew slowly less, the leverage of the cord grew slowly greater and, consequently, the power applied to the wheels was always of the same degree of strength. This invention gave a great impulse to Swiss watchmaking; several centuries later it worked to the disadvantage of English manufacturers, for they continued to use it after other countries had found still better methods of power equalization.

The fusee was invented about the year 1525, at a time when the world was fairly alive with new ideas. People in Europe were just beginning to realize that they were living on a sphere and not upon a flat surface, and that there was a vast new land on the other side of the ocean. Columbus had crossed the Atlantic but a few years before and now explorers were making new voyages of discovery in every direction.

Printing, invented by Gutenberg, about a century before, was becoming common enough to be a real power in the world, bringing the thoughts of men before the eyes of thousands without the slow and expensive process of hand-copying. The first printed copy of the Bible had made its appearance and Caxton had set up his first printing-press all within the lifetime of people then living and printing shops were being established in many places. Many people were learning to read a thing that could be said of very few in the Middle Ages. They were finding out something about the wonderful forgotten civilization of ancient times. Everywhere people's minds were stirring. We call it the time of the Renaissance, or the rebirth of civilization, but in some respects it was more like the awakening of the world after a long sleep. Just as a person on waking looks first at his clock or watch, so now the world, preparing to be busy and modern, needed some better means of telling time. It therefore was both natural and necessary that the watch should have received such a great improvement as the fusee at just this period.

Then began the age of those strange, ingenious watches which we still find in the museums. For some time, there were only a few real improvements. Screws and brass wheels were introduced into their construction about 1550, and glass crystals about 1600. The minute-hand appeared occasionally; but it was not in common use for nearly a century afterward. And that shows how watches were regarded in those days. One would think that such an obvious advantage as that of minute-notation would have been seized upon and utilized at once; on the contrary, people did not seem to care much about it. What was the use of a hand to mark the minutes, when the watch was more likely than not to be half an hour or so in error?

For real timekeeping there were dials everywhere, and there were also fairly good clocks in the towers; at night, watchmen patrolled the streets and called out the hours. These watchmen were the police of the period; it was part of their duty to call out the time, just as the modern police direct people upon the way they wish to go. For timekeeping, the watch was still less useful than the watchman. Made entirely by hand, it was necessarily expensive; therefore, it was made regardless of expense. It was thought of as Malvolio thought of it a possession showing the wealth and station of the wearer, a rich jewel, a toy for noblemen and for kings. Centuries were to pass before real watches were within the reach of common people.

It is said that Edward VI was the first Englishman to possess a watch. This young king, who reigned so short a time, will be remembered by many as the young prince in Mark Twain's famous story The Prince and the Pauper. Mary Queen of Scots had a small watch shaped like a skull a cheerful fashion of the time. Many others were shaped in the form of insects, flowers, animals, and various other objects. Even to-day the Swiss make many watches of curious form.

Queen Elizabeth and her court selected watches as modern women do their hats to match their various costumes. These watches were usually worn on a chain or ribbon round the neck and were largely for display. Several outside cases were often supplied with watches of that period, and they were made to fit on over that which held the works; these were variously ornamented with jewels, tortoise-shell and intricate pierced work in gold, almost as delicate as lace. The covers were decorated with miniature paintings, some of which were very beautiful.

Strangely enough, it was this practise of decorating watches that later gave us our plain white enameled dials, because enamel was the best material on which to paint delicately. To the average museum visitor, the interest in any collection of old watches, aside from their historic association, lies in their marvelously ornamented cases rather than in their mechanism. And in this he very closely repeats the feeling of their original makers or owners; it was more important to follow fashion than to know the time.

This custom of watch-decoration continued more or less through the eighteenth century, and even into the nineteenth, although, by that time, watches had, as we shall see, become excellent timepieces. The story is told that when Dresden was captured by the Prussians in 1757, they found in the wardrobe of Count Bruhl, the Saxon Minister, a different suit of clothes for every day in the year; each had a watch, stick, and snuff-box, appropriately decorated, as part of each one.

Shakespeare never regarded a watch seriously. In Love's Labour's Lost he compares a woman to

A German clock,

Still a-repairing, ever out of frame,

And never going aright, being a watch

A century after Shakespeare's day, Doctor Johnson remarked that a dictionary was like a watch: "The worst is better than none, and the best cannot be expected to go quite true." And Pope says in the same vein:

'Tis with our judgments as our watches none

Go just alike, yet each believes his own.

"Out of a little iron, Peter Henlein constructs clocks which ... can be carried in the pocket." so wrote Johannes Coeleus, in 1511.

All of this reminds one of Dickens' famous character, Cap'n Cuttle, whose watch was evidently of the old school. Readers of Dombey and Son may remember how "the Captain drew Walter into a corner, and with a great effort, that made his face very red, pulled up the silver watch, which was so big and so tight in his pocket that it came out like a bung. "Wal'r," said the Captain, handing it over and shaking him heartily by the hand, "a parting gift, my lad. Put it back half an hour every morning and another quarter toward afternoon and it's a watch that'll do you credit.""

The old idea of regarding the watch as a trinket rather than as a timepiece, as an expensive toy rather than as an accurate and necessary mechanism, has come down to us from the days when a watch was ornamented outside, because it could not be really useful within. Even now, in spite of the modern demand for accurate timekeeping, that attitude has not entirely died away, as is shown by the expression "gold watch" and "silver watch." Of course, there are really no such things; there are merely gold and silver cases for steel, brass and nickel watches. Some people still continue this mistaken idea by thinking of a watch merely as jewelry, as a thing meant more for ornament than for use.

HOW A MECHANICAL TOY BECAME A
SCIENTIFIC TIMEPIECE

Now, since we are at last well into the story of the watch, let us glance back over the road we have traveled. We have seen man first beginning to think of time by noting the positions of shadows or the motions of the stars. Next, we have seen him making his plans for days ahead by means of the changes in the moon, then by making such division in the flow of time as the month, the season, and the year. We have seen him growing out of his savage isolated life in caves and forests and forming tribes and settlements, and have seen him coming out of the darkness of those early ages into Mesopotamia, the Land Between the Rivers, where our first written history seems to begin.

Here, with great cities, temples, and a high degree of civilization and culture, we have found priests studying the stars and making sun-dials and clepsydræ in order to tell the time by shadows, sunbeams, or the dropping of water. We have taken a glimpse at the wonderful people of Greece and Rome, and have seen how, as they became more cultured, they found it necessary to have more accurate means of telling time. We have considered the advantages and disadvantages of the sand-glass, have found clumsy pieces of clock-work in church towers, getting their running power from weights, in order to strike the bells, and have stood with young Galileo in the Cathedral at Pisa, when a swinging lamp gave him the idea of the pendulum.

Lastly, we have seen the making of smaller clocks that were made smaller and smaller until they could be carried as watches, in which springs were used instead of weights. Following this, it has been merely a question of improvement, as one inventor after another has hit upon some idea that would do away with this or that difficulty.

Thus we have come, in the time of Shakespeare, to a clever little contrivance that ticked beautifully but registered time rather badly; that took a long while to manufacture by hand, and cost so much that only the rich could afford to buy it, and that, in consequence, people were proud to own, but did not take seriously as a timepiece.

In all this journey, covering thousands of years, one thing has made itself clear to us the story of timepieces is not a mere mechanical story; it is a human story. Men did not put together certain pieces of wood or metal in order merely to make mechanism, but to meet a vital need. One might almost say that the story of the watch is in the watch itself. The works run and the hands move because of the mainspring, which by pressing steadily forces them into motion. In very much the same way, the busy brains of the inventors and the busy hands of the workmen have been kept active because advancing civilization has been like a great mainspring, always pressing upon larger affairs and greater numbers of people, always needing to fit its engagements more and more closely together, and always calling for better and better means for telling time. Thus, if the watch in the days of Shakespeare and Queen Elizabeth was still an inaccurate timepiece, its improvement was a foregone conclusion. Brains and hands were still active; civilization was still pressing.

It is said that a hog helped in the next development; he helped quite unconsciously by furnishing a bristle. In order to understand this, we must remember Galileo's swinging lamp and the pendulum that the Englishman, Hooke, and the Hollander, Huyghens, applied in the making of clocks. It will be recalled that a pendulum swings in arcs of different lengths in exactly the same time and that this property is called isochronism. Both Hooke and Huyghens could see that the application of isochronism would be quite as valuable in a watch as in a clock, but they realized that this could not be accomplished by means of the pendulum. Therefore, each began to experiment, and each seems to have hit upon the same idea as a substitute for the pendulum in about the year 1665.

This is where the hog's bristle came into use. One end was made fast while the other was bent back and forth by the balance, as it swung to and fro. Being short and stiff, it acted as a spring; in fact, its motion was something like the swing of a small pendulum, and some people incorrectly claim that the name of hair-spring first came from this use of a hair. Of course, a very fine steel was soon substituted for the bristle. Next, it was realized that there would be an advantage if a much longer spring were used, and obviously the only way in which this could be done was by making it in the form of a coil, and so we have the delicate, coiled hair-spring, as it is found in our own watches to-day.

The principle of the hair-spring is not unlike that of the pendulum: the farther the pendulum is swung out from the lowest point of its arc, the greater is the force that gets it back; and the farther a spring is bent from its position of rest, the greater is the force exerted to get it back. With both of these devices it is possible to obtain regular beats and steady motion.

It is hard to realize that nearly a hundred years must have passed by before the hair-spring came into common use. To-day any new device is described in catalogs, written up in the papers, manufactured in quantities and is quickly carried by travelers into every country, but in those days everything was still made by hand, piece by piece, and there was comparatively little travel that would admit of its distribution. Ideas made their way very slowly. In fact, Julien Le Roy rediscovered the principle of isochronism and announced it with a good deal of pride, quite ignorant of the fact that Hooke and Huyghens explained it nearly a century before. And so the hair-spring was slowly adopted by English watchmakers with a number of minor improvements.

Other inventors, of whom presently we shall hear more, worked out better methods of escapement, and the watch movement developed slowly toward its present form. It became possible to tell time more accurately and to make arrangements and plans more closely as the watch became a better time-keeper. The pace of life was speeding up, and people were realizing the value of minutes even of seconds. Therefore the minute- and second-hands were added to the hour-hand that so long had moved alone around the watch-dial. And in 1704, Nicholas Facio, a Swiss doing business in London, introduced jeweled bearings into the mechanism.

The importance of jewels is often misunderstood even at the present day. Many people do not know why jewels are used in a watch, assuming that they are intended for ornament or in some way to increase the value. But most of the jewels in a watch-movement are placed out of sight; and, although they often consist of real rubies or sapphires, they are so tiny and their intrinsic value so small that no watch requires more than one dollar's worth of jewels. They are strictly utilitarian in their purpose. A pivot or bearing, running in a hole drilled in a jewel, creates almost no friction and requires so little oil that a single drop as big as a pinhead is enough for an entire watch. Because jewels are so hard and smooth, a watch with jeweled bearings runs better and wears less and requires less power to drive it, than one in which they are lacking.

During all the time recounted, the great mainspring of civilization had been pressing, ever pressing. Nothing could be considered "good enough" if a way could be found to improve it.

At last an improvement came out of the sea. Travel had been reaching out in every direction; ships were fitted out by scores to take goods from England or the continent of Europe to lands across the seas and to bring back the products of these countries.

The time had been, but a few generations earlier, when people had stood on the shores of the ocean and had wondered what might lie beyond their sight. That water stretched out to the "edge of the world" they felt sure, but what there happened to it they could not tell. Surely, however, it must be peopled with monsters and demons. It was foolhardy to venture too far from land. We can hardly realize what a piece of insane rashness it must have seemed to most people when Columbus sailed out boldly into this vast mystery, nor how the world was thrilled when he brought back word of strange lands and strange peoples he had found beyond the horizon.

But by the time now reached in our story the oceans had become highways of trade, and men were beginning to draw those strange, crude maps of the continents, which make us smile until we stop to think how maps might have looked had they been left for us to make. At all events, the problems involved in navigation were being much discussed in every land.

One of the greatest of these problems was to discover the whereabouts of the ship at any given time. When one is out of sight of land the sense of location necessarily becomes inoperative; one wave looks like another, and there are winds and currents which might carry a ship hundreds of miles out of its course unless there were some way of knowing its true position. At first, the stars, and later the compass gave help in giving direction but not in showing position. How might this be done? There was no possible way in which the element of telling time did not enter.

That sounds a bit strange until one stops to think of the rotation of the earth once in twenty-four hours. If one could travel around the earth, from east to west, at a uniform rate in exactly twenty-four hours, he would find clocks and watches indicating the exact minute he started at every step of his journey; and the sun would remain steadily at the same height above the horizon, if he always kept to one parallel of latitude. His rate of speed would have to be about eighteen miles a minute, if he chose to travel along the equator, or to state this same thing in another way, when it is noon in New York, it is 11 A. M. in Chicago, 10 A. M. in Denver and 9 A. M. in San Francisco; it is also 1 P. M. several hundred miles out into the Atlantic; 2 P. M. still farther out; 5 P. M. in London; and so on. In other words, it is some one of all the moments of the twenty-four-hour day at the same time, but the time that indicates each of these moments is different at different points. Therefore, if you could find out the time at any point, and could compare it with the time at the place you had left, you would know just how far east or west you had come, but not how far north or south.

Ascertaining the time was not difficult; at noon it would be shown by the sun. Nor was it difficult to compare the time provided one had an accurate timepiece, but a watch that ran either fast or slow might mislead one by hundreds of miles. You can see how important it was that navigators have some means of exactly measuring time. This was one of the points at which the great mainspring of civilization pressed hardest upon the brains of inventors and the hands of workmen.

So, from the sixteenth century onward, the leading governments of Europe offered large rewards for a chronometer sufficiently accurate to determine longitude at sea. In England, Parliament offered twenty thousand pounds, or one hundred thousand dollars, for a time-keeper which, throughout a voyage to the West Indies, would give the longitude within thirty miles. This meant that it must keep time within a minute a month, or two seconds a day. Both Huyghens and Hooke somewhat naively attempted to make a pendulum clock keep time at sea; but imagine the action of a pendulum while a ship was rolling and tossing!

The problem was really one for the watchmaker, since a clock is made for keeping time while standing in one position and a watch for keeping time while being moved about. John Harrison, the inventor of the famous gridiron pendulum, finally won the munificent prize. In 1762, after several trials and failures, he succeeded in producing a timepiece which varied, under test, only a minute and four seconds during a voyage of some five months. This was excellent timekeeping far within half a second a day; it made it possible for a captain at sea to determine his position within eighteen miles. Harrison's mechanism was too complicated for description in these pages. Indeed, it was so difficult of comprehension that, before paying him his reward, the English government asked Harrison to write a book of explanation in order that his inventions might be copied by other makers. He did so and finally received the money. Harrison's ideas have now been greatly simplified, but, in general, his plan is used in the making of marine chronometers to this day; thus, in a sense, it is due to Harrison's brain that our great ships are able to cross the ocean on almost schedule time.

Both the first success of the chronometer and the later efforts toward improving it had a great influence upon the next few generations of watchmakers; the final improvements were made in the days of the American Revolution. It was at this latter period that a man named Thomas Mudge worked out the kind of escapement that is still used in our watches. A little later, the Swiss-Parisian, Abraham Louis Breguét, improved the hair-spring by bending its outer coil across the others to their center and fastening it at that point in order that the spiral of the spring should expand equally in all directions from the center.

The last development of importance consisted in doing away with the fusee. The faults of this device had been the need of a thick watch to give it room, and the danger that a broken mainspring might destroy other parts of the movement in its recoil. French and Swiss watchmakers reduced the friction until it needed very little power to run the mechanism, and then were able to employ a mainspring which was not stiff enough to require a fusee.

American makers adopted this idea, but the British clung to the fusee and the stiff spring; it has cost them much of their prestige as watchmakers and much of their trade.

Thus, the mechanism of both clocks and watches was practically in its present state by the year 1800. The "grandfather's clock" of that date may look old-fashioned, but it tells time a modern way, and the mechanical ideas in George Washington's watch were not so very different from those which we find in our own. There have been many small improvements since, but the great inventions had all been made.

It is interesting to remember that most of these inventions are due to the English artisans of the seventeenth and eighteenth centuries, although in delicate workmanship and beautiful decoration, they were equaled and perhaps excelled by the Swiss and by the French. The work of producing a satisfactory timekeeping machine, begun by priests and by astronomers, and carried forward by the demands of the navigator and the patient labor of the craftsman, had ended after thousands of years, in triumph. The ticking contrivance of wheels, levers, and springs was no longer a mechanical toy; it was a marvelous instrument which was made by man with his head and hands and yet was almost as accurate in its action as the sun and stars themselves.

Here ends the first great division of our story. The scientific problem had been solved; what remained was to democratize the keeping of time; to place mechanism equal to the best of those days within the reach and within the means of every man. In this later development the work was to pass out of the hands of artists and inventors into those of manufacturers. Its history from this point on is no longer a record of science but a romance of industry.

THE WORSHIPFUL COMPANY AND ENGLISH WATCHMAKING

From the beginning, there are two sides to the history of timekeeping. The first is the story of discovery and invention how men labored for thousands of years to produce a contrivance that would really tell the time. But if only a few such machines existed in the world, it would be of very little use to humanity in general, however perfect each might be. Accordingly history must now recount how clocks and watches came to be made in sufficiently large numbers and at sufficiently low cost to be within the reach of all who needed them.

The turning-point from the inventive to the industrial side of the development was reached about the year 1800. Timekeeping has always been a part of history, and history a part of timekeeping, and this opening of the nineteenth century was a period when history itself was changing, for the progress of civilization is like a journey over a mountain road; one must needs turn occasionally or one can rise no higher. The American Revolution had ended but a few years before, and the thinly settled states were trying the strange experiment of having the people govern themselves without a king. In the old world, the people of France had suddenly risen up and seized the power from their king, and a bloody struggle had ensued in which many of the old nobility had been beheaded. In England, the power of the throne was growing less and the power of the people greater. In fact, the whole world was becoming more and more filled with democratic ideas and ideals than ever before.

Now, this same democratic idea that set up republics was getting ready to put a watch into every man's pocket. At first, everyone had told the time for himself, and had told it badly. Now, after thousands of years, it had come about that a few had the means of telling time accurately. The great inventors mentioned in the last few chapters had contributed one idea after another, until, among them all they had worked out clocks and watches that would keep correct time. But these timepieces were not yet convenient in form, and they certainly were not yet convenient in price for the average man. They still were made by hand in small quantities, and such a condition would have to be changed before it would be possible for everyone to tell the time and to tell it well.

Naturally, the industrial and business development of watchmaking began long before 1800, long before, indeed, the time at which the inventions were all complete. For centuries the two sides of the story, the inventive and the industrial, had progressed side by side, but for the sake of clearness, we have described the inventions first. Now we must glance back again to the time of Shakespeare, when the period of modern inventions was just beginning, in order to see how the business side of watchmaking started upon its growth.

Four nations have been concerned in this development England, France, Switzerland, and the United States. The English worked in one way; the French worked in another; the Swiss, in still another; while the Americans took up the final organization of the work in a manner that was thoroughly typical of their peculiar genius.

The mechanical improvements and inventions were mostly made, as we know, by the English. But for the beginnings of the watch industry in England one must go back to a time before the days of Hooke and Huyghens, to the year 1627, the year of incorporation of the Worshipful Clock-makers, Company. Imagine such a name being chosen to-day! The Worshipful Clock-makers' Company was the original trade-organization of the business in England. It was not at all like our modern companies but was one of those great trade "guilds" which played such an important part in the development of European industry.

London about 1600
London about 1600
Octagonal Rock
Crystal Watch
Octagonal Rock Crystal Watch
French, 1560-90
Square French Watch
Late Sixteenth Century
Square French Watch
Late Sixteenth Century
Oval French Watch 1590
Oval French Watch
1590

Shell Shaped Rock Crystal Watch, French
Shell Shaped Rock Crystal Watch
French
Cross Shaped Rock Crystal Watch, French
Cross Shaped Rock Crystal Watch
French
Book Shaped Swiss Watch, 1560-1600
Book Shaped Swiss Watch
1560-1600
When Watches Were Jewels
Watches of the Sixteenth Century, with but one hand, and pierced metal or rock crystal cases. In the collections of the Metropolitan Museum.

People sometimes think of the medieval trade-guild as something like the modern trade-union, but this is a mistake; it was in many ways quite different. Perhaps one might call it a sort of a cross between a labor-union and a manufacturing trust. Within a certain district, all who were occupied in a particular business were required to belong to the guild; otherwise they were not allowed to do business, and the "district" might include the whole country. In order to gain an idea of a guild, imagine in this country a single association of jewelers to which everyone connected with the jewelry business was forced to belong, whether he were manufacturer or retailer, employer, or employee, the head of his firm or the last new clerk behind the counter. Or, to look at it in another way, imagine a trust controlling the whole industry and a union including all the workmen under a closed-shop system, and then suppose that the trust and the union were one and the same. That would be like one of the great medieval guilds. It was easy for such an organization to create a monopoly of the entire national product.

Sometimes the guild would forbid the importation of foreign goods and would not permit workmen to come from other countries. It usually regulated, to some extent, the conditions of wages and labor. It fixed its own standards of quality of the product; if goods did not come up to this standard, they might not be sold, and the rules of the guild had practically the force of law. But it did not attempt to control prices, nor to limit the quantity of production, nor to interfere, except very indirectly, with free competition among its own members.

Thus, it was not, in our modern sense of the conception, a company at all, but an association of independent manufacturers or tradesmen, each in business for himself, each in competition with his fellow craftsmen, and all kept upon a tolerably even footing by limiting the amount of labor that each one might employ. Its members were the master craftsmen, each the head of his own house; through them were associated the journeymen, or skilled workmen in their employ, and the apprentices. These latter might rise to be masters, in business for themselves. But no one without such a connection could engage in the business at all, in any capacity whatever.

The Worshipful Clock-makers' Company, under its charter granted by Charles I, had the power to make rules for the government of all persons following the trade within ten miles of London, and for regulating the trade throughout the kingdom. Its first master, or president, was David Ramsay, who was mentioned as having been "constructor of horologes to His Most Sacred Majesty, James I," and is one of the characters in Scott's novel "The Fortunes of Nigel." Its wardens or executives were Henry Archer, John Willowe, and Sampson Shelton; and there was, besides, a fellowship, or board of directors. The company proceeded at once to forbid all persons "making, buying, selling, transporting, and importing any bad, deceitful clocks, watches, larums, sun-dials or cases for the said trade," and full power to search for, confiscate and destroy all such inferior goods, "or cause them to be amended."

This company limited the volume of business by forbidding any one master to employ more than two apprentices at one time without express permission; and, since all journeymen must first pass through the stage of apprenticeship, this tended to keep up wages by limiting the labor supply and to keep competition on a fair basis. The coat of arms of the company represented a clock surmounted by a crown, the feet resting upon the backs of four lions, all of gold, upon a black ground; on either side were the figures of Father Time and of a king in royal robes; and the motto beneath read: Tempus Imperator Rerum, or "Time, the Emperor of Things." These matters sound rather quaint to us, but perhaps the quaintest of them all is the idea of a monopoly concerning itself so jealously with the quality of the product, and letting prices and competition practically alone.

It was under such conditions that the English work was done and the inventions made. Huyghens was, of course, not an Englishman; and Hooke was rather an inventor and a scientist than a manufacturer. Both these men themselves made clocks and watches, but they made them only as instruments to assist them in their researches, or as working-models of their design. It was often said of Hooke that he never cared to develop an invention after he had proved that it would work. But once these first inventions had been adopted, the real production of timepieces was in the hands of the Clock-makers' Company, and the great names were those of clock-makers.

These were the days when the leaders of the industry worked with their own hands as well as with their heads. We may imagine the master seated in the front room of his shop studying over a new model, or putting together and decorating one already made; or, perhaps, making with his own hands some of the most delicate parts. From the back rooms would come the sound of tapping or filing as the journeymen and apprentices were hard at work upon their various tasks. Meanwhile, perhaps some apprentice, standing outside the door, would call out to passers-by and urge them to step in and buy. This was a favorite form of advertising in that time. For that matter, we still have our "barkers" and "pullers-in" at Coney Island and elsewhere. Everything about the small business was carried out under the personal direction of the master and, where necessary, by his own hand. The phrase "clockmaker to the King" meant something more when applied to such a man than merely that royalty had purchased some product of his craft.

Such a one was Thomas Tompion, often called "the father of English watchmaking." He was the leader of his craft in the time of Charles II and he, more than anyone else, worked out the inventions of Hooke for actual manufacture. He left his father's blacksmith shop to become a clock-maker, from this he went on to the more delicate work of making watches, and at last became a famous master of his guild. It may fairly be said of him that he set the time for history in his day, for most of the royalty and great men of Europe timed all their doings from banquets to battles by Tompion watches.

Meanwhile, he, too, was making watchmaking history by his improvements. Tompion made watches with hair-springs, balance-wheels and escapements with various improvements. His design of the regulator is nearly that in modern use. His cases, too, were as famous as the movements that he made. The so-called "pendulum watches" were then much in fashion, and Tompion met the demand by making a number of them. They did not, of course, work with a pendulum; but one arm of the old foliot balance could be seen through an opening in the case or dial, and looked like a pendulum swinging to and fro. To read the advertisements of that day one would think that all lost or stolen watches were of Tompion's making, so often does his name appear in them.

Many legendary stories are told about Tompion's work. It has been set down in cold print that Queen Mary gave one of his watches to Philip II of Spain, and that he made watches for Queen Elizabeth. Unfortunately for such stories, Tompion was not born until 1638, by which time both Mary and Elizabeth had been dead for some years. But though the legends themselves are untrue, yet they do shed some light upon their subject, for such stories, true or false, are not told about unimportant men. And it is true that Tompion grew so celebrated that at his death, in 1713, he was buried in Westminster Abbey, where only the great may have resting-places. Another famous watchmaker was George Graham, the inventor of the mercury pendulum. He first was Tompion's journeyman, then his partner, and at last became a well-known astronomer, having become interested in astronomy through making astronomical clocks. But his great contribution was the invention of the dead-beat escapement, which, in one form or another, is in use in all the best clocks and watches of the present time, and which has had more to do with making their accuracy possible than has any other improvement since the discovery of the isochronism of the pendulum and hair-springs. Graham, also, is buried in Westminster Abbey; his body lies beside that of Tompion, his teacher and friend.

Another famous figure was Daniel Quare, the first to devise the mechanism for driving the two hands as we have it to-day. Quare was a Quaker, and was no less prominent in the Society of Friends than in his business. As a Quaker, he was opposed to taking an oath of any kind, and was what we now call a "conscientious objector" to warfare. Therefore, at the same time that he was being honored by royalty for his work, he was being prosecuted and fined for his refusal to pay taxes for the support of the army and of the Established Church. When he was made clock-maker to King George I, means had to be devised for excusing him from taking the oath of allegiance.

It was Quare who originated the practise of giving to each watch a serial number, so that it could always be identified. This is, of course, a common custom with us; we also number automobiles, and many other manufactured articles of value, and Quare's device of numbering watch-movements may very well have given the start to all this.

Still other famous watchmakers were Harrison and Arnold and Earnshaw, who between them developed and perfected the marine chronometer that we discussed in the last chapter; and Mudge, in whose hands watch-movements really became modern in type. Men of this kind thought first of producing reliable work which would give service; ornaments, curiosities of workmanship, and even convenience, were secondary. Some of these men were extremely independent; for example, Arnold, in his early days and by way of establishing a reputation, made a repeating watch less than a half-inch in diameter so small that it was worn set in a ring; but when King George III had bought the masterpiece, and the Empress of Russia offered one thousand guineas (more than five thousand dollars) for a duplicate, Arnold coolly excused himself on the plea that he desired the specimen to remain unique.

Time passed; machinery began to be employed in manufacturing and hand-work declined. The guild system in every line slowly changed into our modern organized industry. This was only natural, for factories were becoming larger, their output was increasing and the head of the business was no longer likely to be himself a master workman. The greater part of this change, of course, took place in the nineteenth century, and was primarily owing to the increased use of machine-power and improvement in transportation. But as regards watchmaking in England, the substitution never became complete, for the bulldog quality in the Englishman has always made him hold fast to his ideas. Habits died hard, and the old methods were changed slowly and under protest, even when these changes spelled progress.

At first, as we have seen, the watch was the work of one man and of his assistants, and was almost entirely handmade. In those days, the trade was supplied by a multitude of small independent manufacturers. To make a single watch might take weeks or months; and every one must be made separately and patiently, regardless of labor or expense. So long as this method could hold its own, the English watchmakers led the world; their watches were good, but they certainly were not cheap.

The gallant of Colonial times often carried two watches, as was the fashion, but often they were both unreliable.

After a time, other countries began to use more modern methods, and English watches could no longer stand competition in the world's markets. However, the bulldog quality still held; English manufacturers preferred to lose ground rather than change their methods. The introduction of machinery and the employment of women operatives were each bitterly opposed. Factory production was never adopted on a large scale, nor was there much combination of small independent manufacturers. Necessarily, these things did, at last, come to be done; but half-heartedly, and without much success. At one time, for example, there were some forty small factories making various parts which each watch manufacturer assembled and adjusted for himself.

The Clock-makers' Company is still in existence; although now, of course, it has developed into a society like the ordinary modern association of manufacturers. Under pressure of change and competition, English manufacturers were compelled unwillingly to change their system of production, but the character of the watches they would not change. The same country which had made so many of the mechanical inventions finally settled down into satisfaction with its models at a time when other nations were continuing to make improvements, as, for example, when they clung to the fusee after watchmakers abroad had found a better substitute.

The English watch has remained heavy, substantial, and reliable; it is an excellent mechanism produced regardless of expense. Such a watch cannot be made cheaply, least of all by British methods. There has been something obstinate in the maker's attitude; if the law of supply and demand called for something different, so much the worse for the law. The English have been slow to see the possibilities in the cheap watch. They have not realized that a watch need not be expensive in order to keep good time. They started to put the watch into universal use, but left to other nations the completion of the process.

WHAT HAPPENED IN FRANCE AND SWITZERLAND

Across the English Channel lives a race of a very different character. The French are people of highly adaptable minds; often they see possibilities in the inventions of other nations which those other nations have failed themselves to see. The automobile was first made in the United States, but the French soon developed it into something that was better than our early clumsy cars, and we were years in overtaking them. The Wright Brothers first learned the secret of aerial flight, and then Wilbur Wright sailed for France, where the people went wild with enthusiasm over the idea of flying; it was in France that aviation really became what it is to-day.

The French have always been fine mechanics and finished workmen. It was to be expected that they would do something artistic and interesting with the manufacture of timepieces. They could not make a better watch than the British were turning out toward the end of the eighteenth century. Nobody could but they could make it more beautiful. In Shakespeare's time and afterward, while watches were still more valuable as works of art than they could be as timepieces, the richest work of this nature was done in France. There watches were made in the form of mandolins and other musical instruments, in the form of flowers, in the form of jeweled butterflies, and in wonderful cases, painted and enameled and engraved. In the J. Pierpont Morgan collection in the Metropolitan Museum of Art, New York, there is a watch which, in 1800, on the fete-day after the battle of Marengo, Napoleon Bonaparte gave to Murat, who was his brother-in-law and one of his generals. On the back cover of this watch appears a miniature portrait of Napoleon himself. And since he himself was the author of the gift, one may assume that it represented the Great Emperor's own conception of himself.

The wrist-watch, to-day a military necessity, was at first a French idea. It is interesting to learn that the merchants and makers of this kind of work were in their own time called neither watchmakers nor horologists, but toymen. There again is shown the old idea about watches; they were not timepieces but toys.

Later on, toward the end of the period of invention, when first, the clock, and soon afterward, the watch, had become fairly accurate timekeepers, the French makers again took the lead in the same way; once more they beautified what they could not practically improve. The French clocks of the period of Louis XIV and his successors are celebrated for their design. One might easily suppose, from an examination of the great modern collections of rare and precious watches in our museums that the French had been the leading watchmakers of the world, for the specimens there found being selected chiefly for beauty or value from the collector's point of view, are oftener of French than of any other make. Yet it must not be supposed that the French made no inventions. The credit for some of the important improvements is disputed between the English, French and Swiss, and it is not always easy to decide which nation has the better claim. Furthermore, certain of the French watchmakers came from Switzerland while at various times, some of those in France moved to England, especially during the reign of Terror. The distinctions are somewhat confused and we can only speak in a general way.

However, while the watchmaking industry was developing in France, it gave forth a seed which took root in new soil. In the hill country of eastern France, in the town of Autun, there lived a watchmaker named Charles Cusin. One day, in 1574, for reasons that we do not know, he moved a few miles eastward across the border into Switzerland and there settled in the beautiful lake city of Geneva. He probably had no thought that this personal act of a private citizen would have an effect upon history, but an industry employing thousands of people and making millions of dollars worth of goods can be traced back to the time when he crossed the border.

Remember that this was back in the days of Shakespeare and Queen Elizabeth, while watches were still esteemed jewels and ornaments for the wealthy, and when the improvements which later made them practically useful had not yet been invented. The business side of watchmaking was thus growing up at the same time with the inventive and scientific; it was preparing itself for the day when the mechanism should be perfected, and the only remaining task would be to popularize its perfection.

Charles Cusin liked Switzerland and thirteen years later he became a citizen. In the course of time, he was active in founding a watchmaker's guild in Geneva and from that period Geneva watches have been famous. This does not mean that Switzerland had contained no watchmakers before Cusin's appearance, but we are considering the beginnings of a great industry and not mere instances of isolated workmen. The man from Autun seems to have been one of those energetic leaders who see possibilities and know how to organize. It is largely through such men that the world progresses.

You will remember that in an early chapter we touched upon the way in which men first began to exchange the results of their work in order that each man might devote most of his time to the special task for which he was best fitted, such as hunting, or the making of weapons. Through this exchange, everyone was enabled to live better than anyone could have lived by himself. But if it were true that people doing different things could help each other, it also became true, after a while, that people doing the same thing could help each other and could help the general public, by learning to co-operate. They could exchange ideas, improve their work, and bring about better conditions. This was one of the effects of the guilds they changed crafts into industries.

The guild with which Charles Cusin now had to do some say he was its sole founder was a very dignified and important board of master-workmen. It was founded about fifty years earlier than was the Worshipful Clock-makers' Company in England, and its members were no ordinary workmen. Switzerland was, and still is, a thoroughly independent little country and a man skilful enough to make a whole watch with his own hands was apt to be a man who realized his own worth.

The members of this guild were decidedly particular about their dignity and their meetings were serious occasions, as may be seen from Article I of their regulations which read: "Whenever the master workmen shall meet in a body to discuss subjects pertaining to their guild, they shall, before proceeding to such discussion, offer prayer to God beseeching Him that all that they say and do may rebound to His glory and may further the interests of these people."

As a matter of fact this dignity was based upon a correct conception that has been somewhat overlooked in the present busy age. The man who has to do either with the manufacture or sale of timepieces does well to take his position seriously since he is a most important link in our entire civilization. Such a man may well reflect upon the fact that without the timepieces which he produces or sells, the world would drop into hopeless confusion, for human society is able to run smoothly and efficiently only when it is correctly timed. Workmen and dealers engaged in such a vital industry have a great responsibility to their fellow-men.

It is probable that members of this guild who met from time to time in the Swiss city by the lake shores, under the shadows of the snow-topped Alps, realized something of this responsibility. Their timepieces were not yet as accurate as are ours of to-day, and the world was not yet so busy that its affairs required the closest adjustment, but they at least were trying earnestly to keep the human cogs running smoothly by turning out watches as nearly perfect as their skill and knowledge would permit.

This may be seen again in Article V of their regulations; "The functions of the jurors are to enforce the laws of the guild and to provide that there be no infringement of the same. To this end, they shall be required to visit each journeyman at least four times during the year, having power to seize all articles which do not conform to the specifications now in force, to report all delinquents to the worthy governing board, and to punish the offenders in accordance with the gravity of their fault."

Grew more elaborate and ornamental, but scarcely more useful.

In the collections of the Metropolitan Museum.

It is quite clear that Geneva was out for quality in watches, and, indeed the name of the Swiss city has always been associated with quality. Nevertheless, they were no angels those old Swiss craftsmen; they were in fact quite preponderatingly human. Thus it was not long before they began to make a tight little monopoly of their business. They restricted the number of workmen who might be admitted to the guild, and they secured special ordinances by means of which all other watchmakers were forbidden to establish themselves within a certain distance of the city. In other words, they did not purpose allowing the new and promising industry to grow beyond their control.

There were, however, other independent people in those days who hadn't the slightest intention of being bound by such restrictions. Here and there, a watchmaker left Geneva to carry on his work in some foreign city, as, for example, in Besancon, France. Thus began a competition which grew and spread as time went on.

This competition developed some interesting features. For example, the guild in Geneva obtained the passage of laws forbidding anyone from bringing into the city, in a finished state, a watch constructed within a certain distance. "Schemes" for watches and certain parts might be made at will, but only members of the citizen guild were permitted to complete these schemes.

Such restrictions naturally did not tend toward low-priced watches; but all watches in those days were necessarily high-priced, and a man wealthy enough to afford one was apt to seek the best that could be bought. Geneva's strictness gave it so great a reputation that during the seventeenth and eighteenth centuries foreign watchmakers flocked to the Swiss city very much as art students later journeyed to Paris, and it became the acknowledged center of the European industry. As time went on the demand for time-pieces became more widespread and many Genevans moved to other cities where they became dealers in Geneva watches. It is said that, in 1725, the city of Constantinople contained as many as eighty-eight mercantile agents who had become established in this way.

One hundred years after the founding of the guild, Geneva was producing five thousand watches a year, having one hundred masters of the guild and three hundred journeymen. Now five thousand watches is no small output when it is considered that each one must be constructed entirely by hand and occupied a matter of weeks in the making; yet, by 1799, the city contained nearly six thousand watchmakers and jewelers and was producing fifty thousand timepieces a year.

Not many miles to the northward from Geneva is another mountain city that of Neuchatel. Neuchatel also contained an enterprising and skilful population, for the Swiss people seem to have been naturally ingenious and skilful in the use of tools. Doubtless the mountainous character of the country has had something to do with this fact; farming and fruit-raising are slow, hard work in their rocky soil and severe climate and the making of bulky articles is not desirable where transportation must be had over mountain trails.

The Swiss with their clever fingers had long been famous for their wood-carving; now, when they had a chance at an industry which called for delicate and skilful hand-work and which produced goods of small size and high value, it exactly suited them.

Geneva "saw it first," but kept it so closely to herself that it was several generations later before watches were known in the Neuchatel district not far away, yet, this district is another great center of the industry.

It is said that in 1680, more than one hundred years after Charles Cusin moved to Geneva, a horse-dealer from the little town of La Sagne, came home from his travels and brought with him an English watch. Great was the wonder that it excited among the simple people of his native place. They passed from hand to hand the little ticking mechanism which had the strange power to tell time, and then one day the ticking ceased, which perhaps is not surprising, in view of the freedom with which the watch had been handled.

The horse-dealer knew nothing of the mechanism but was very anxious to have the works set right. It chanced that there was a young locksmith in La Sagne, a lad of only fifteen, named Daniel Jean Richard, who was so skilful and ingenious that he had already made repairs in the tower clock of the village. "Show the watch to Daniel Jean Richard" said everybody.

The delighted lad began to take the delicate mechanism apart, studying carefully each wheel and spring and lever until he felt that he understood exactly how it should work. Then, when he had succeeded in reassembling the parts and in making the watch tick bravely once more, he was seized with a great ambition to build another one all by himself.

After many experiments with his crude locksmith tools, he did produce a watch which would run and which would tell time after a fashion the first watch ever made in the Neuchatel district but it did not satisfy his artist's soul and he realized that he must have better tools.

Somebody told him that there was in Geneva a machine for cutting wheels, and he set out to see it for himself, only to come back sadly disappointed. Wherever he asked to see the machine, the canny Geneva craftsmen shook their heads. This eager lad from another town had far too intelligent a face to be allowed to learn the precious secrets. The most that they would do was to let him have a few of the wheels made by the machine.

Then he began to work out for himself a machine to cut the wheels, and at last succeeded in the task, so that before long he was well on the way to becoming a watch manufacturer. Richard, however, was generous with his ideas; he instructed a number of the young men of his district, so that watchmaking soon began to flourish in his town and in those about it.

We have now seen how the watchmaking industry became established in two great centers in Geneva, where the highest quality was maintained, but under the rule of the guild, which did not encourage quantity of output, and in the Neuchatel region where no guild system existed. In the course of time this latter region overtook and passed in quantity of output that of Geneva. By 1818, the Neuchatel district of the Jura was turning out watches at the rate of 130,000 a year.

The solid old Geneva watchmakers criticized their rivals as being less exacting in quality and less careful as to the standard of gold used in their cases, but the Neuchatel people had no difficulty in finding customers; we read that one hundred and forty of their merchants went twice a year to the Leipsig fair, where they sometimes sold watches to the value of four million francs ($800,000) in a year.

The two principal centers of Swiss watchmaking have been mentioned although, of course, watches were made in other districts as well. It is easy to see that many generations ago it had already become a very large industry, and so we need not be surprised to learn that even to-day the tiny inland country produces a larger annual export value of watches than even our vast United States. Watchmaking has been so large a source of wealth that the Swiss government has aided it in every way, including the establishment of schools and courses for training skilled workmen. More than sixty thousand Swiss people are directly employed in the Swiss watch industry and over three hundred thousand, or one-twelfth of the entire population, are indirectly connected with it. The Swiss have also made many inventions and improvements so that they have had much to do with the development of the watch itself as well as with the industry.

As we have already seen, it was a Swiss who invented the fusee, another who introduced the use of jewels for reducing friction and the stemwind is also of Swiss origin. It was the Swiss, too, who, early in the nineteenth century, did away with the solid upper plate which covered the works and used, instead, a system of bridges. The bridge form of movement allows each part to be repaired or adjusted separately and to-day it is to be found in all watches of the higher grades.

The Swiss invention of the fusee, described in Chapter VIII, played an important part for several hundred years, but at last it was replaced by something simpler and still more effective. Made to equalize the difference in the pressure exerted by a stiff mainspring when first wound up and when partly run down, it worked beautifully but was rather clumsy; and it required comparatively heavier parts which naturally necessitated the use of greater power. Thus friction and, consequently, wear were increased. But the Swiss by making watch-parts that were very light but yet strong, and by reducing friction principally through the introduction of jewels into the mechanism, succeeded at last in getting a movement that could be run with very little power. So they now could use a weak and slender mainspring, made so long that only its middle part ever was wound and unwound, and thus the pressure remained equal, and the use of the fusee was no longer necessary. This principle, called the "going barrel" construction, reduced friction, and made the thin modern watch a possibility. The American makers, as we shall presently see, adopted the "going barrel" construction practically from the first. They had no traditional prejudices, and they knew a good mechanical idea when they saw it.

But the British would have none of it. Their national bulldog quality set its teeth on the old idea that had given them their heavy, substantial, accurate watches, and hung on grimly. The Swiss watches might be lighter and more graceful but they questioned their lasting qualities. The Swiss could make watches more beautifully, but the English were suspicious of cheapness and declined to adopt the new development.

Thus the English, who up to about 1840, had led the world in the manufacture and sale of watches, began to fall behind. The American watch industry was then in its infancy, and the French industry had never been of any great size. The Swiss gradually drew ahead until they practically gained control of the world's market for watches. Switzerland became known as the place from which watches came, and, very much as "Havana" stands for a fine cigar, so a fine watch was apt to be called a "Geneva."

In former days, Swiss workmen made some particular watch part in their own homes, while so-called "manufacturers" bought the parts and "assembled" the watches.

This, then, was the situation at about the middle of the nineteenth century when watchmaking in America was beginning to grow into a large industry. The French had always made good watches and very beautiful and elaborate ones too, but they never made very many. The English were falling behind so far that it was said, in 1870, that half the watchmakers' tools in England were in pawn. The Swiss were in control of the business, making both the best and the worst watches in the world and by far the greatest number.

Everywhere a good watch was still too costly to be owned by anyone of moderate means, while cheap watches were little more than toys which could not be depended upon either to wear well or to keep good time.

In spite of all developments, therefore, there still remained the need both for a high-grade watch at a reasonable price and for a cheap watch that would be accurate under rough usage. These things were genuinely necessary, for the world was growing steadily away from the theory of special privilege, and the requirements of the average man were becoming more insistent.

From those early days, when the astrologers in Mesopotamia had kept their knowledge a secret for themselves, down through more than forty centuries, only a few had possessed the means of accurately telling time; but now had come the railroad, the telegraph, the modern factory, the newspaper and many other developments which speeded up the movements of humanity in the rush and whirl of modern life until it had become absolutely necessary that the means of measuring and performing those movements in an economical manner should be within the reach of every man.

It remains to be shown how American watchmaking discovered this need and organized to meet it; how it found and filled the gap that had been left in foreign watchmaking, between high-priced watches that were good, and low-priced watches that were not good; how it developed a cheaper good watch and a better low-priced one than the world had so far known; and how, in so doing, the American industry has grown within the memory of living men to such an extent as to take second place, and, in many respects, first place in watchmaking throughout the world.

HOW AN AMERICAN INDUSTRY CAME ON HORSEBACK

At last the clock industry came to America, and it came on horseback. If you had been upon a dusty country road in Connecticut about the year 1800, you might have seen a plainly dressed young man come riding along with a clock strapped to each side of his saddle and a third fastened crosswise behind him.

"Hello, Eli Terry!" you might have heard some farmer sing out, as the rider drew near.

"Hello, Silas," the other would call back; "don't you think it's about time you bought a clock?"

"Can't afford it, Eli; it takes me a long time to make forty dollars raising wheat."

"Yes; but you can't afford to be without one, Silas." And, dismounting, he would unstrap one of the clocks and bring it up to the stone wall. Then would follow the period of bargaining, so dear to the shrewd, hard-headed sons of Connecticut. Perhaps when young Terry climbed back into the saddle and said "Gid-dap," one of his clocks would stay behind with the farmer. Like most successful salesmen, Terry was a close observer of human nature; he knew that habits once formed are hard to break. He discovered early that if a prospective customer could be made to depend upon a clock for telling time, the clock would soon sell itself. One day, during a rain-storm, he sought refuge in a farmer's home. He brought in with him one of his clocks and placed it on the mantel over the fireplace, explaining that he would like to leave it there, where it would not get wet, while he continued on his journey.

"I'll be back for it in a few days," he said, as he waved good-by.

When Terry returned, some days later, the farmer realized that the clock, which he had first regarded as an extravagance had somehow become a necessity, and, with no urging on Terry's part, the sale was quickly completed.

Some of the original clocks are still running in the very farmhouses where Eli Terry succeeded in selling them, and where they have ticked off the minutes of American history since the days of Adams and Jefferson. They were truly remarkable clocks, in spite of the fact that their works were cut out of hard wood with country tools, and put together by a carpenter.

The first American clocks were made of wood, and most of the early clockmakers were at first carpenters. We have seen clockmakers developing from priests and astronomers and blacksmiths and locksmiths and jewelers; but here is a new gateway to the trade. This came about naturally enough in a country where the cheapest and most plentiful material was wood, and where the carpenter and joiner was accustomed to constructing every possible thing of it. Eli Terry of Connecticut was one of the best known of these early New England craftsmen. He was born in East Windsor, just a few years before the Revolution. By the time that he was twenty, he had made a few clocks, cutting the wheels out of hard wood with saw and file, and making wooden hands, dials, and cases. Then he moved to Plymouth, not far from Waterbury, and set up a small shop where he employed several workmen. They would make a dozen or two at a time, entirely by hand. Then Terry would take these out and sell them, sometimes as far as the "new country" across the New York state line.

It took a long time to make a clock in this way, even for fingers that were as clever as Terry's, and it is no wonder that he was compelled to charge from twenty to forty dollars apiece, a sum, which, by-the-way, would be equal to at least four times as much to-day according to the difference in the purchasing power of money. We must remember, too, that a family then bought its clock as it bought a wagon or a spinning-wheel, almost as a man buys his house to-day. Certainly it was a far more important transaction relatively than the purchase of a motor-car.

Probably, if one could have overheard some of these roadside clock-sales it would have been noted that the bargaining was not all upon one side, for there was not a great deal of money in circulation, and people were very apt to "swap." Likely as not, Terry would have to take his payment in lumber, in clothing, or in some other commodity and these, in turn, he would dispose of when an opportunity presented itself. This was more or less the type of the old horseback Yankee trader of the days when men still remembered the Revolutionary War. These were the days when a man who produced some one thing might be forced, in order to realize on its value, to trade it for almost anything else.

When we think of the early American timepiece, we generally picture to ourselves the so-called "Grandfather's Clock," the kind with the tall case which Longfellow wrote about as standing on a turning in the stair and ticking away: "Forever!" "Never!" "Never!" "Forever!" as it marked the passage of the years. But Eli Terry, the first of all American clock-makers, could not well carry such a big contrivance with him on his horseback trips; therefore, while he made the works for these clocks, he left it for other people to construct the cases; the clocks which he sold complete were those which could stand upon a shelf or hang upon the wall.

After a time, his orders increased to a point where he felt justified in moving into an old water-power mill and rigging machinery to do some parts of the work. Thus we find machinery used in American clock-making almost from the beginning of the industry. Terry thus was a real manufacturer; he had grasped the importance of machine production in contrast to hand-craftsmanship.

The move paid; it cut the cost of making nearly in half and greatly increased the output. He now could afford to sell his clocks more cheaply, and the business grew at once. After a while he began to make clocks in lots of one or two hundred and then, indeed, his neighbors shook their heads gravely. "You are losing your mind, Eli," they told him, in solemn warning. "The first thing you know, the country will be so full of clocks that there will be no market for them. You are getting reckless and ruining your business."

But Eli Terry followed his own judgment instead of that of the croakers; before he died he was making ten or twelve thousand clocks in a year and was selling them too. They brought him a fortune.

Thus was the industry of making timepieces born in America. It began in New England, which is still the chief center of manufacture, and it began with clocks, not watches, for the simple reason that in those days, a watch was a luxury whereas a clock was a necessity. Like the watch industry in Switzerland, American clock-making was an active business from the start, and, as we have seen, the man with whom it started was a typically Yankee combination of ingenious mind, skilful fingers, and a knack for business.

Of course, the conditions of life in America at that time had a great deal to do with methods used in building up the industry. Instead of a civilization centuries old that had wealth, rank, royalty, and a complete organization of all methods of living, here was a new country learning to do things in its own way.

It is hard for us to imagine the conditions which prevailed when our whole population was a mere fringe of scattered settlements along the Atlantic seaboard; when people made long trips on horseback or by stage-coach and men wore powdered wigs and knickerbockers; when New York was a small town on the lower end of Manhattan Island, and Chicago had not even been dreamed of. Still, it was necessary to tell time, and our thrifty ancestors needs must watch the minutes in order to save them as thriftily as they saved everything else. Not one person out of hundreds, in a country where a living must be wrung from the soil by means of hard work, could afford to own anything so expensive as a watch, but every one felt it necessary to have a clock, if possible, and it became one of the greatest treasures of the home.

Eli Terry, America's first clock manufacturer, peddled his wares among the shrewd, hard-headed sons of Connecticut.

This, then, was the market in which Terry and those who followed him had to sell. It was a market that could not afford to pay for ornament but desired practical service at low cost. What was needed, therefore, was a clock that would keep time and cost not a cent more than was absolutely necessary. The American industry was forced to start upon a basis entirely different from that of Europe.

As Eli Terry's business grew, he needed assistance, and he secured the help of a young mechanic named Seth Thomas from West Haven, and the two worked together for some time.

The name of Seth Thomas has appeared upon so many clock-dials that it is perhaps the best known name in all American clock-making. He was a good mechanic, and a good business man, and he had ideas of his own about increasing trade. In the course of time, about the year 1800, he and a man named Silas Hoadley bought the original Terry factory in the old mill, and set up business for themselves. Terry, however, established himself elsewhere and continued to manufacture clocks.

Thus the industry was growing; there were now two factories instead of one. Seth Thomas prospered by adopting each popular fashion or improvement in clocks as it came along and applying it upon as large a scale and as honestly and well as could be done. He built up such a reputation that even to-day, while the name of Seth Thomas on a clock face does not suggest any particular form or style of clock, it is associated with good time keeping and honest workmanship.

The third of the famous old New England clock-makers was Chauncey Jerome. He was a man younger than Terry and Thomas by nearly a generation. Like both of his predecessors he was brought up to the carpenter's trade, and like both of them he was a born New England trader. But of the three, Jerome was perhaps most the inventor and least the man of business. As a boy, he worked for Seth Thomas when Thomas was still building barns and houses. He worked for Eli Terry in the old shop at Plymouth. Then, after a period of soldiering in the War of 1812, he went back to clock-making, sometimes manufacturing by himself and sometimes associated with one or the other of the two older men, or in other firms and enterprises too numerous to follow. Always he seems to have been somewhat of a rolling stone, although in his time he gathered as much moss as the best of them: always he was inclined to experiment with new ideas.

Jerome's carpentering skill caused him to be first interested in the making of cases, and most of the familiar forms of old American clocks the square clock with pillars at the corners and a scroll top, the clock with a mirror underneath the dial and the like, were designed by Terry and Jerome between them. Later on, when the establishment of brass foundries in Waterbury and Bristol had enabled American makers to construct their work of brass instead of wood, Jerome worked out a design for a brass one-day timepiece in a wooden case, small enough for easy transportation, and cheaper than any clock ever made up to that time. Its price at first, near the place of manufacture, was only five or six dollars, but afterwards this was reduced.

This low-priced clock was as remarkable in its way as was the dollar watch, which it foreshadowed. And like the watch, it would not have been possible except through machine work and quantity production. It was a success at once and Jerome's business rapidly increased. In 1840, he was established in Bristol, turning out the new clocks by the thousand, and rapidly making a fortune. A year or two later, he decided to send a consignment of them to England.

Again, people shook their heads and prophesied failure. "You're losing your mind, Chauncey," they told him as they had told Eli Terry before him.

The older wooden movements could not, of course, endure a sea voyage without swelling and becoming useless. A brass movement could, of course, be sent anywhere, and some of the more expensive ones had been shipped to all parts of the country, yet it seemed absurd enough to send American clocks to England where labor was so cheap to England, which was then the chief clockmaker of the world. Nevertheless, Jerome persevered, and his son sailed for London with a cargo of the cheap clocks. At first, the English trade would have none of them. No clock so cheap could possibly be good, they said, and Connecticut was the home of "the wooden nutmegs." It was only after great difficulty that they were introduced. Young Jerome got rid of the first few by leaving them about in retail stores, asking no payment for them until sold.

The enterprise was saved by an event which was a joke in itself. The English revenue law at that time permitted the owner of imported goods to fix their taxable value. But the government could take any such property upon payment of a sum ten per cent greater than the owner's valuation. Jerome's clocks were valued at their wholesale price, and were presently seized by the customs officials on the ground that this valuation was fraudulently low.

The elder Jerome chuckled upon learning of this. He was well satisfied to have closed out his first cargo at ten per cent profit, and at once sent over another shipment which was taken over by the customs as promptly as the first. But by the time the third consignment arrived, enough of the clocks had been sold to establish a demand for them among the retailers, and the officials finally conceded that the low price might be a reasonable one after all.

Jerome was not at the height of his prosperity. He had the largest and probably the most profitable clock business in the country; and, in the few years following, his product was exported to all parts of the world. Then the Bristol factory burned down and he moved to New Haven, where the Jerome Manufacturing Company enjoyed a brief period of great success. The business was constantly extended, and the wholesale price of the cheap brass clocks was brought as low as seventy-five cents. This figure seems almost impossibly low for the time, but the authority for it is Jerome's own autobiography.

A few years before the Civil War, the Jerome Company failed and, curiously enough, this failure came about through its connection with that usually successful man, P. T. Barnum, the famous showman. The story is too much complicated to be given here in detail, but it seems that Barnum had become heavily interested in a smaller clock company, which was merged with the Jerome concern. The overvaluation of its stock, combined with mismanagement and speculation among the officials of the Jerome Company, served to drive the whole business into bankruptcy. Barnum lost heavily, and it took him years to clear up his obligations. Jerome never did recover from it; after some years of failing power in the employ of other manufacturers, he died in comparative poverty.

His long and eventful life spans the whole growth of the American clock business from the days of Eli Terry and his handsawed wooden movements down to the maturity of the modern business supplying, by factory methods and the use of specialized machinery, millions of clocks to all parts of the world. He had made clocks all over Connecticut, in Plymouth, Farmington, Bristol, New Haven and Waterbury, as well as in Massachusetts and, for a time, in South Carolina and Virginia. He had worked with his hands for Terry and Seth Thomas at the old wooden wheels and veneered cases, which were peddled about the country and sold for thirty or forty dollars each to be the treasured timekeepers of many households. And he had headed a modern factory, turning out dollar clocks by the tens of thousands.

It is said that a child in the first few years of its life lives briefly through the whole evolution of civilized mankind. That "infant industry," American clock-making, likewise, in the short space of fifty years passed through most of the steps of the whole growth of time-recording between the Middle Ages and our own era. This country stands now among the leading clock-making nations of the world; its product is famous in every land and a timepiece from Waterbury or New Haven may mark the minutes in the town from which Gerbert was banished for sorcery because he made a time-machine, or in that land between the rivers where the Babylonians first looked out upon the stars.

Most of the American clocks are still made in Connecticut; in fact, more than eighty per cent of the whole world's supply (excluding the German) comes from the Naugatuck Valley. The New Haven Clock Company, which is the successor of the Jerome Company, is to-day one of the largest. As far back as 1860, it was producing some two hundred thousand clocks a year. The Seth Thomas Company and others of the historic concerns are still at work in various portions of the state. And the Benedict and Burnham Company, with which, at one time, Chauncey Jerome was associated, became the Waterbury Clock Company, now regarded as the largest clock producer, and of which we shall hear more later on.

The key-note of the whole development was that new principle which American invention, prompted and stimulated by the pressing necessities of a new nation, brought into the business of time-recording the principle of marvelously cheapening production-costs without loss of efficiency, through the systematic employment of machinery on a large scale.

As long as the inventive brains and the technical knowledge of the old-time craftsman found expression only through his own fingers, the results would be limited to his individual production, and the costs would be proportionately high. When, however, the master mind was able to operate through rows of machines, each under the supervision of a mechanic trained to its particular function, his inventive genius was provided with ten thousand hands and a hundred thousand fingers. Furthermore, the production gained in quality as well as in quantity, because of specialization, all the time its costs were in process of reduction. This, perhaps, has been America's chief contribution, not only to the making of timepieces, but, also to the world's industry in general.

These huge but beautiful clocks represent the most reliable form of timepiece known to the people of the Seventeenth and Eighteenth Centuries. In the Metropolitan Museum.

AMERICA LEARNS TO MAKE WATCHES

While Eli Terry was sawing wood for his curious clocks back in the early days of the nineteenth century, Luther Goddard, America's first watch-manufacturer, was preaching the Gospel to the town and country-folk in Massachusetts and Connecticut. Between sermons he repaired watches.

Although we can find no record of such a meeting, it is easy to imagine that while plodding along some dusty country road Preacher Goddard met Terry jogging along with his cumbersome wooden clocks hanging from his saddle. The thought may have come to the minister-mechanic that it would be much easier to peddle watches than clocks.

Whatever may have been the prompting, we find, as a matter of record, that, in the year 1809, while Terry was making and peddling his clocks, Luther Goddard set up a small watch-making shop in Shrewsbury, Massachusetts, the place of his birth. He employed watch-makers who had learned their trade in England. At that time, there was a law in force which prohibited the importation of foreign-made watches into America and this gave Goddard his chance. But in 1815, when the law was repealed and the American market was quickly flooded with cheaper, if not better watches from abroad, he was forced to retire from the field. During those few years he had produced about five hundred watches.

Discouraged by his venture into worldly affairs, he turned again to his former occupation of preacher and evangelist, and consoled himself with the remark that he "had here a profession high above his secular vocation." In those days, protection and free trade had not yet become the rival rallying cries of two great political parties; otherwise we might have found this early manufacturer entering politics instead of the pulpit. While he is credited with manufacturing the first American watches, however, it is doubtful whether he and his workmen really did more than to assemble imported parts.

More than twenty years now passed before another effort was made to produce watches in America this time by two brothers Henry and James F. Pitkin of Hartford, Connecticut. In 1838, they brought out a watch, most of the parts of which were made by machinery, but it proved more or less a failure. After a brief struggle, they gave up in discouragement. Henry Pitkin died in 1845, and his brother, a few years later.

While the Pitkin Brothers were struggling with their problem in Hartford, Jacob D. Custer of Norristown, Pennsylvania, was engaged in a similar task. He succeeded in making a few watches between 1840 and 1845, thus gaining his niche in history as the third American watch manufacturer.

But all of these were merely forerunners, for now there stepped upon the stage a young man whose ability and perseverance were destined to launch American watch-making fairly upon its way. This young man was born in Hingham, Massachusetts, in 1813, and his name was Edward Howard; it was born in him to be an inventive and ingenious craftsman and to feel toward the mechanism of time-keeping the devotion of an artist to his art. At the age of sixteen, he was apprenticed to Aaron Willard, Jr., of Roxbury, one of the cleverest clock-makers of his time.

Young Howard took to clock-making as naturally as a Gloucester man takes to the sea. Some of the clocks he then made are still ticking as vigorously as ever. Having presently learned all he cared to know about clock-making, he cast about for other fields of action. His bent, as he himself said, "was all for the finer and more delicate mechanism," and it was natural that these qualities of the watch should absorb his interest. It was equally natural, since he was an American clock-maker at a time when that trade was being revolutionized by machine-work, that he should dream of applying such methods to the watch.

"One difficulty I found," he is quoted as saying, "was that watch-making did not exist in the United States as an industry. There were watchmakers, so-called, at that time, and there are great numbers of the same kind now, but they never made a watch; their business being only to clean and repair. I knew from experience that there was no proper system employed in making watches. The work was all done by hand. Now, hand-work is superior in many of the arts because it allows variation according to the individuality of the worker. But in the exquisitely fine wheels and screws and pinions that make up the parts of a watch, the less variation the better. Some of these parts are so fine as to be almost invisible to the naked eye. A variation of one five-thousandths of an inch would throw the watch out altogether, or make it useless as a timepiece. As I say, all of these minute parts were laboriously cut and filed out by hand, so it will readily be understood that in watches purporting to be of the same size and of the same makers, there are no two alike, and there was no interchangeability of parts. Consequently it was 'cut and try'. A great deal of time was wasted and many imperfections resulted."

Howard's ambition lay in the production of a perfect watch for its own sake; and he wanted to make it by machinery, believing that, in that way, it could be made most perfectly. Other people had thought of the same thing. Pitkin had attempted it, and there had been some experiments of like nature in Switzerland. But the man who loves his work as Howard did will succeed in anything short of the impossible, because neither time nor labor, neither failure nor discouragement, matter at all to him as against the hope of making his dream come true.

As Howard was emerging into young manhood, the great period of American invention was rapidly developing. Morse was struggling with the electric telegraph which he invented and perfected in 1835, and Goodyear was busy with machinery and processes for enabling rubber to be used commercially, thus laying the foundation for one of the greatest American industries of to-day. Ingenuity was in the air and invention was conquering realms that had been believed beyond reach.

When people told Howard that it was absurd to think of improving upon the manual skill of centuries, he answered that he expected to make his machinery by hand. And when they said that a machine for watch-making would be more wonderful than the watch itself, he only laughed and agreed that this might be so.

To-day, we are familiar with such phrases as "standardized parts" and "quantity production," which explain to us how it is possible for a single factory to produce millions of watches in a year, or for another kind of plant to turn out half a million automobiles in a like period. The way in which "quantity production" came about is curiously interesting. Watch-making received one of its greatest impulses from a famous American inventor who probably would have been amazed had anyone told him that his idea upon quite another subject would some day help to put watches into millions of pockets.

There is no particular connection between a cotton-gin and the "quantity production" of watches, but it is interesting to know that the same ingenious brain which designed the one also unconsciously suggested the other. Late in the eighteenth century, Eli Whitney gained lasting fame as the inventor of a machine which would automatically separate the seeds from the fiber of crude cotton a machine which revolutionized the cotton industry of the south.

In 1798, Whitney secured a contract to manufacture rifles for the government. He decided that they could be made much more rapidly and cheaply if he could find some way to produce all the separate parts in large quantities by machinery, and then merely assemble the various parts into the completed weapon. The inventive mind which was capable of devising the cotton-gin found this new problem to be comparatively simple, and it was not long before Whitney was making thousands of rifles from machine-made "standardized parts," where only one could be made before. Half a century later his machinery was still turning out rifles parts in the great arsenal at Springfield, Massachusetts, and it was not until this period that it exerted a distinct influence upon watch-making.

While Howard in Roxbury was dreaming of producing watches by machinery, another young man Aaron L. Dennison, of Boston was also obsessed with the same dream and grappling with the same problem. It is therefore not strange that the paths of these two soon crossed. Born in Freeport, Maine, 1812, Dennison was just a year older than Howard. He was an expert watch-repairer and watch-assembler, having learned his craft among the Swiss and the English workmen in New York and Boston. The year 1845 found him conducting a small watch and jewelry business in Boston.

Some few years earlier, Dennison had visited friends in Springfield, Massachusetts, and while there he was taken to one of the interesting show-places of the town the Springfield Arsenal. As he made his slow progress through the great rifle factory, he marveled at the wonderful machinery and the system which had originated in the brain of Eli Whitney nearly half a century before; Whitney was dead and gone, but his works still lived.

Dennison returned to Boston, fired with an ambition to apply the Whitney system and methods of rifle-making to the manufacture of watches. He brooded over the scheme for years, constructing a pasteboard model of his imaginary watch factory and planning in detail its organization.

Then occurred a meeting that was to make history a meeting marking the first step in founding a great American industry and wresting from Europe and Great Britain the watch-making monopoly which they had continuously held since the days of the "Nuremburg Egg." Dennison met Howard, and the contact of the two minds was like the meeting of flint and steel. Dennison shared Howard's belief that watch-parts could be made better and more accurately by the use of machines. He had the watch-making experience and Howard the mechanical skill to design the new machinery. One may imagine how the two young men inspired each other. They had the ideas; all they now needed was the capital and this was supplied in 1848 by Mr. Samuel Curtis, who backed them to the extent of twenty thousand dollars.

Dennison immediately went abroad to study methods in England and Switzerland and came back more than ever convinced of the soundness of their own ideas.

"I have examined," said he, "watches made by a man whose reputation at this moment is far beyond that of any other watchmaker in Great Britain and have found in them such workmanship as I should blush to have it supposed had passed from under my hands in our own lower grade of work. Of course I do not mean to say that there is not work in these watches of the highest grade possible, but errors do creep in and are allowed to pass the hands of competent examiners. And it needs but slight acquaintance with our art to discover that the lower grade of foreign watches are hardly as mechanically correct in their construction as a common wheelbarrow."

Reached the extreme of elaboration and costliness, but were not always equally successful as time keepers. In the collections of the Metropolitan Museum.

On his return, in 1850, he and Howard established themselves in a small factory in Roxbury, under the name of the American Horologe Company. And that little factory was the foundation of what is now the great establishment of the Waltham Watch Company, the first and hence the oldest watch company in America, and the parent concern of most of the rest.

It was perhaps at this time that an employee, one P. S. Bartlett, returned to his home town on a visit and was asked by his old neighbors what he had been doing.

"I am working," said he, "for a company which makes seven complete watches in a day." Great was the merriment at this reply. "Why, where on earth could you sell seven watches a day?" they shouted.

With the advent of the factory, the real troubles of Dennison and Howard began. It is worth while to glance for a moment at the problem which lay before them, if only to appreciate its difficulty. The old plan was to have a model watch made by hand by a master workman. This watch was then taken apart and its separate parts distributed for reproduction by a multitude of specialized workers involving perhaps some forty or fifty minor trades. These parts, hand-made after a hand-made model, were then returned to the expert who assembled and adjusted them. At the worst, this resulted in gross error; at the best, in individual variation. A part from one watch could not be expected to fit and work accurately in another, although the two were supposed to be alike in all their parts.

The new idea was first to lay out the whole design on paper and then to make the various parts by machinery according to the exact design. It was supposed that a machine making one part would duplicate that part repeatedly without variation; that in so far as the machines themselves were accurate, the parts produced would necessarily be interchangeable; that any set of parts could therefore be assembled without fitting or alteration. The finished watch, it was assumed, would require adjustment only. Theoretically, this idea was correct; practically, it could not be perfectly carried out, and the results did not fulfil the hopes of the manufacturers. In the first place, there were not in existence any machines of the required delicacy and precision; every one must first be invented, then designed, then made, and finally adjusted for practical operation. Even so, and notwithstanding the great mechanical achievements of the Waltham Company, the results never succeeded in realizing the dreams of Howard and Dennison, of absolute interchangeability of parts. It remained for the Ingersoll organization, many years later, to develop such a factory system.

Before Howard and Dennison could make a single watch, therefore, they had to invent all the mechanism, and themselves build and install every invention. Moreover, several of the processes had to be worked out from the ground up. There was nobody in America who understood watch-gilding, for example, or who could make dials or jewels.

Thus they set to work developing the machinery as fast as they could do so, and imported such parts as they themselves could not yet make. It was a staggering task and a discouraging devourer of capital. "I do not think," said Dennison many years later, "there were seven times in the seven years we were together that we had money enough to pay all our employees at the time their wages were due. Very often we would find ourselves without any cash on hand, but Mr. Howard would manage some way to produce enough to tide over with."

The two men made a perfect team, eager to give each other credit, and each having unbounded loyalty and confidence in the other and in their enterprise. But, curiously enough, it was Howard, the artist and dreamer, who seems to have developed into the business man of the two, in addition to being the inventor and engineer, whereas Dennison, the expert watch-repairer, became the designer and originator of plans. It was said of him long afterward that there was probably never an idea in American watch-making that had not at some time passed through Mr. Dennison's resourceful mind. He is known to many as the "Father of the American Watch Industry," although he insisted that Howard deserved the title as much if not more than he. Dennison schemed out what was to be done, while Howard found the money and invented the machinery with which to do it.

Their first model, an eight-day watch, was Dennison's idea. It was found to be impracticable and was soon abandoned in favor of a one-day model. The name of the company had to be changed, because it did not find favor with some of the English firms from whom they bought certain parts. They called it the "Warren Manufacturing Company" for a time, and their first few watches were marked with this name. Later on, they moved to a new factory at Waltham and incorporated under the name of the Waltham Improvement Company. It was while the act for its incorporation was before the Massachusetts legislature that some wag there produced the couplet:

"A Waltham' 'patent' watch, which ere it goes

Besides the 'hands' must have the 'ayes' and 'noes."

All this time, the tools and machinery were giving trouble. There were innumerable difficulties. For example, New England workmen objected to cutting the pinion-leaves because they were shaped like a bishop's miter. And financial pressure was always upon them. The building was one of the earliest attempts at concrete construction, and was far from stable in stormy weather. Mr. Hull, afterward foreman in the dial-room, said: "Often in those days we would jump from our stools when we felt something jar, for fear the building would fall down. Somehow, it never did."

In 1854 the name was changed again, this time to the American Watch Company. Incidentally, Mr. Dennison took his place among the large and honorable company of inventors who have been called insane. He earned that title by saying that they would eventually make as many as fifty watches a day. The company now makes between two thousand and three thousand a day.

Just as they were on the point of a richly deserved success, the panic of 1857 drove the young company into bankruptcy. The plant was purchased by Royal E. Robbins, of the firm of Robbins and Appleton, watch importers. Howard went back to the old factory at Roxbury, taking with him a few trained workmen, and patiently started all over again. He succeeded, at last, in producing really fine watches, although in small numbers; and his new business, as we shall see later, developed into the E. Howard Clock Company, and practically abandoned the manufacture of watches. Meanwhile, the Waltham factory, under good business management and with Dennison as its superintendent, was safely steered past the financial rocks and shoals of the period, and began gradually to reap the reward of its less fortunate early efforts.

It was the Civil War, with its great military demand for watches, which first set the Waltham Company squarely upon its feet by justifying quantity production. A dividend of five per cent was declared in 1860; and one of one hundred and fifty per cent in 1866, the short-lived Nashua Watch Company having meanwhile been absorbed. Since that date its name has been twice changed first, to the American Waltham Watch Company, and then to the Waltham Watch Company, which is now its title.

At the present day, the Waltham Company employs nearly four thousand people and produces about sixty-eight thousand complete watch-movements a month, or over three-quarters of a million a year.

This output is made possible only through the extensive employment of automatic machines, all of which have been invented and manufactured at the Waltham factory. Even now it is not possible to buy watch-making machinery ready-made in the open market; it is all "special" work, designed and often built by the watch manufacturers themselves. And the development of this great industry, employing, at first, crude devices operated for the most part by hand-power, to the complex automatic mechanism which seems to act almost with human intelligence, has been a marvelous achievement.

The company now makes ten different sizes of regular movements, in more than a hundred different grades and styles. Of these every part is made in the Waltham factory. It was the first establishment in the world in which all parts of a watch were made by machinery and under the same roof. And its success revolutionized the methods of watch-making not only in America but, to a less degree, in all parts of the world. A prominent London watchmaker who went through the plant in the early period of its success said to his colleagues: "On leaving the factory, I felt that the manufacture of watches on the old plan was gone." And the name passed into literature when Emerson, describing a successful type of man, said, "He is put together like a Waltham watch."

CHECKERED HISTORY

One of those mental marvels who can play fifteen simultaneous games of chess, blindfolded, might be able to form a complete idea of the American watch-making industry in the years that followed the Civil War; all that the ordinary mind can gain is a bewildering impression of change and confusion, with companies springing up, and merging or disappearing, all over the industrial map. Inventions were as thick as blackberries in August and, to investors, as thorny as their stems. Countless revolutionary ideas in watch-making revolved briefly few evolved, and capitalists, large and small, learned the sobering lessons of experience, as capitalists ever have and ever will.

When P. S. Bartlett boasted that his company was making seven watches a day, his friends laughed, "Why, where could you sell seven watches a day?"

With it all, certain points seem to stand out as clearly defined among them the fact that watch-production appealed strongly to the public mind at a time when the nation, galvanized into intense activity by the great conflict, was entering an era of extraordinary self-organization. This is, of course, significant. The nation's time as well as its forests, mines, and other resources, must be a factor in the growth of public wealth, and this could not be unless it were widely and accurately measured, which, in turn, implied the universal use of the watch.

The later history of American watch-making is, therefore, a story of the formation of many companies, the failure of most, and survival in the case of comparatively few. In the sense of being founded by men whose experience had been gained at Waltham, the Waltham Company was more or less the parent of the majority. Of the failures, it may roughly and broadly be stated that the general trouble was most often a lack of cooperation between technical watch-making skill and business management.

Of the occasional successes due, on the other hand, to perfect harmony between these two factors, the Elgin National Watch Company, established at Elgin, Illinois, in 1864, was one of the first. Its officials and promoters were not watchmakers but business men a group of Western capitalists who organized the company at the suggestion of a few trained men from Waltham, to whose technical experience and knowledge they gave entire liberty of action from the first. This combination of Western enterprise and Eastern mechanical skill was a great and immediate success. Within six years from its incorporation, the Elgin Company had built its factory, designed and made its own machinery, and marketed forty-two thousand watches. It is said to be the only American watch company which has paid dividends from the beginning. And yet this achievement cannot be traced to anything strikingly distinctive either in the policy or in the product. It was a case of doing rapidly and easily, with vast previous experience to build upon, what the parent company had so long strived to accomplish, and of doing this honestly and well. In a small way, it was like the rapid growth of democratic principles in America, having, as it were, the British commonwealth of a thousand years on which to base itself.

The period of the development of American watch-making was also the period of the rapid and enormous expansion of railroads. The two were naturally related, in that railroading demands the constant use of a great number of watches, while its progress in punctuality and speed is in direct proportion to the supply of reliable timekeepers. Precision is here the great essential; every passenger must have the means of being on hand in time in order not to miss his train. But what is of far greater importance, railroad men must know and keep the exact time not alone for their own protection but in order that they may protect and safeguard the lives of those who are entrusted to their care.

Most of our great inventions and improvements can be traced to some pressing human need. Many of them, unfortunately, are delayed until some great catastrophe shows the need. It required a disastrous wreck to bring home to the railroads and make clear the necessity for absolute accuracy in the timepieces of their employees.

In the year 1891 two trains on the Lake Shore Railroad met in head-on collision near Kipton, Ohio, killing the two engineers and several railway mail-clerks. In the investigation which followed, it was disclosed that the watches of the engineers differed by four minutes. The watch which was at fault had always been accurate and so its owner took it for granted that it always would be. But tiny particles of dust and soot find ways of seeping into the most carefully protected works of a watch, and every watch should be examined and cleaned occasionally. So it was with the engineer's watch. A speck of coal dust, perhaps, had caused his watch to stop for a few minutes and then the jolting of the engine had probably started it running again. That little speck of dust and those few lost minutes cost human lives. This wreck occurred not many miles from Cleveland, Ohio, then and now the home of Webb C. Ball, a jeweler, who as a watch expert, was a witness in the investigation which followed. His interest thus aroused, he worked out a plan which provided for a rigid and continuous system of railroad watch inspection. The plan which he then proposed is now in operation on practically every railroad in the country.

A railroad watch must keep accurate time within thirty seconds a week, and is likely to be condemned if its variation exceeds that amount in a month; it must conform to certain specifications of design and workmanship which are only put into movements of a fairly high grade. And the railroad man must provide himself with such a timepiece and maintain it in proper condition, subject to frequent and regular inspection by the railroad's official inspector. There is thus a compulsory demand for watches of a definite quality and performance at a reasonable price.

Expressly to meet this, the Hamilton Watch Company, of Lancaster, Pennsylvania, was organized in 1892, the year after the wreck which started this reform. This company therefore represents an enterprise founded for a specific purpose and concentrating upon a certain specialized demand, although this does not mean that it is the only company which caters to the needs of the railroad man. All of the great companies produce timekeepers of the highest precision for railroad use, but the Hamilton Company has devoted itself more particularly to supplying this one field.

The Gruen Watch Company, of Cincinnati, Ohio, is typical of still another line of endeavor the beautifying and refining of watch-cases and watch-works. Its founder, Dietrich Gruen, was a Swiss master watchmaker. He came to America, as a young man, in 1876, married here, and established the international industry which bears his name. It might be said that his watch is not an American product, as the Gruen movements are made at Madre-Biel, in Switzerland, and then sent over to America to be cased, adjusted, and marketed. Perhaps the most notable contribution of this company to the watchmaking industry was to inaugurate the modern thin type of watch. This was evolved by Frederick, the son of Dietrich Gruen, and was made possible by the inverting of the third wheel of the watch, so that the whole train runs in much less space than was previously required.

These four companies are by no means the only successful ones, but they do typify the general trend of development of the American watch industry from 1850 until near the end of the nineteenth century, when a new and even greater era in the history of timekeeping was inaugurated. The story of this development will be considered in later chapters. In the period then closed, however, the ideal of Dennison and Howard, which most people then regarded as an impossibility, was realized to a degree which they themselves would never have thought possible. Dennison died in 1898 and Howard in 1904.

Although watch-making is the creation of European genius and was rooted in European experience, with boundless capital at its command and carried on in communities trained for generations in the craft, it is in this country that it has been brought to its fullest modern development. The census figures, while incomplete and somewhat misleading, are expressive of the amount of growth and of its nature. According to these figures there were in 1869 thirty-seven watch companies in the United States, employing eighteen hundred and sixteen wage earners, or an average of less than fifty workmen; and their combined product was valued at less than three million dollars. In 1914, the last normal year before the Great War, there were but fifteen such companies; the law of the survival of the fittest had been operating. But these fifteen employed an average of over eight hundred people, or twelve thousand three hundred and ninety in all, and the combined value of their product was stated as over fourteen million dollars. These figures are far below reality in that they do not include the large volume of watches produced in clock factories.

American watch-making is typical of the difference between the American and European industry in the nineteenth century. Here a complete watch is produced in one factory, while in England, Switzerland and France most establishments specialize in the manufacture of particular parts and these parts are then assembled in other factories. Some fifty different trades there are working separately to produce the parts. And the manufacturer, whose work is chiefly that of finishing and assembling, takes a large profit for inspection and for the prestige of his name.

By the American system, a thousand watches are produced proportionately more cheaply than a dozen; and a thousand of uniform model more cheaply than a like number of various sizes and designs. Automatic machines tend to economy of labor and uniformity of excellence. The saving begins with the cost of material and ends with the ease and quickness of repairs due to the standardization of parts.

Lord Grimthorpe said: "There can be no doubt that this is the best as well as the cheapest way of making machines which require precision. Although labor is dearer in America than here, their machinery enables them to undersell English watches of the same quality."

It now remained for American ingenuity and enterprise to level the ramparts of special privilege in the world of time-telling by producing an accurate and practical watch in sufficient quantity and at a price so low as to place it within the reach of all.

THE WATCH THAT WOUND FOREVER

The most important development in any affair is naturally the one which concerns the greatest number of people. In the United States, it is the people who count and nothing can be considered wholly American which does not concern the mass of the population. We have already seen how watch-movements were brought to a high degree of accuracy, and have followed some of the steps by which the industry was developed in the United States, but there remained one great step to be taken, and that was the putting of an accurate watch within the financial reach of almost every person. The way in which this was brought about was thoroughly American.

In 1875, Jason R. Hopkins, of Washington, D. C., after many months of patient labor, perfected the model of a watch which he thought could be constructed in quantities for fifty cents each. He secured a patent on his model, and with Edward A. Locke, of Boston, and W. D. Colt, of Washington, sought to interest the Benedict and Burnham Manufacturing Company, of Waterbury, Connecticut, in its manufacture.

Failing in this, Locke abandoned further effort so far as the Hopkins' model was concerned. Hopkins, however, continued, and finally succeeded in enlisting the active support and financial resources of W. B. Fowle, a gentleman of wealth and leisure, who owned a fine estate at Auburndale, Massachusetts. This led to the formation of the Auburndale Watch Company. Within a few years, Fowle had sunk his entire fortune of more than $250,000 in the enterprise, and the Hopkins watch had proved a complete failure. In 1883 both Fowle and the Watch Company made assignments.

There are many who still remember the great Centennial Exposition at Philadelphia in 1876, celebrating the one hundredth anniversary of the declaration of American Independence. Those who were there may recall the interesting exhibit of a huge steam-engine at least, it seemed huge at that time and, in a glass case near by, a tiny engine so tiny that it could be completely covered by a small thimble. This midget steam engine, with its boiler, governor, and pumps, was just as complete in all of its parts as was the big engine. Three drops of water would fill its boiler. It was a striking example of mechanical skill and fineness of workmanship, for it had been made under a watchmaker's microscope with jeweler's tools.

The most interesting thing about this little engine was that, unknown to its designer, it heralded the dawn of Democracy in the Kingdom of Time-telling, just as it then was helping to celebrate the birth of American freedom. In the spring of 1877, Edward A. Locke, of Boston, who two years before, as we have seen, had been interested in the Hopkins' watch, visited the neighboring city of Worcester, and while strolling along the main street, in a leisurely manner, he chanced to glance in the window of a watch-repairer's shop. There he saw the tiny engine which had excited so much wonder and admiration at the Philadelphia exposition the year before. For many months, Locke and his friend George Merritt, of Brooklyn, New York, had been thinking and dreaming of the possibility of supplying the long-felt and rapidly-growing need for a low-priced watch a pocket-timepiece that could be sold for three or four dollars. The cheapest watch in America at that time cost ten or twelve. They had searched in vain for a watchmaker who was ingenious or courageous enough, or both, to attempt the making of such a timepiece.

Fascinated by the marvelous little engine, Locke stepped into the shop and spoke to the lone workman at the bench near the window. This obscure and humble watch repairer was D. A. A. Buck, the proprietor of the shop and designer of the engine, who was soon to gain renown as the inventor of the famous Waterbury watch.

For the sum of one hundred dollars Buck agreed to study the problem, and, if possible, design for Locke a watch which would meet his requirements. Day and night, for many weeks, he labored at this task, and finally submitted a model. It was not satisfactory.

Worn by his labors and disappointed by his failure, he fell ill. Some days later, Mrs. Buck sought out Locke and joyfully told him that her husband had worked out a new design which he believed would correct the defects of the former model and that, as soon as he recovered, he would begin work upon it. Within a few months he had completed a second model. This time he was successful.

Then began the struggle of Locke and his associates to interest capital in the new enterprise. Most of the preliminary funds and factory space were provided by the Benedict and Burnham Manufacturing Company, a brass manufacturing concern at Waterbury, Connecticut, and the predecessor of the present Waterbury Clock Company. Thus the new watch came to be known as the Waterbury.

Within the next twenty-eight months many thousands of dollars had been raised and expended before a single watch could be turned out for sale. It was not until 1880 that the Waterbury Watch Company was finally incorporated and ready for business. Then the factory proudly produced its first thousand watches. They were perfectly good-looking watches, but they had one important weakness they would not run, because, as it was found, the sheets of brass used in stamping out the wheels had an unfortunate grain, and the wheels would not remain true. Another thousand were made with this defect corrected. This time most of the watches would keep time, but there still was a large percentage of "stoppers." After more study, experiment, and expense, the product was improved until only about ten per cent of the watches refused to run, and the Waterbury watch was really on the market.

It was a wonderfully simple piece of mechanism, very different from the ordinary watch. The whole works turned round inside of the case once every hour, carrying the hour-hand with them. The mainspring was coiled round the outside of the movement, so that the case formed a barrel, and was wound by the stem. It had the old duplex escapement of the days of Tompion and the dial was printed on paper, covered with celluloid and glued to the plate. It had only fifty-eight parts, kept time surprisingly well, was not much to look at, but was sold at the then unheard-of low price of four dollars.

It was put on the market with real Yankee ingenuity. Some of us remember when Waterbury watches were given away with suits of clothes, and the pride with which, as youngsters, we exhibited our first watches thus obtained to our playmates who were less fortunate. The nine-foot mainspring required unlimited winding, which was one of its chief joys, and our friends often solicited the privilege of helping in the operation. Some of the more ingenious among us held the corrugated stem against the side of a fence and made the watch wind itself by running along the fence's length, while other children looked on enviously.

In spite of the disadvantage of the time necessary for winding, perhaps in part because of it, the Waterbury watch became famous the world over and reached a very large sale for its day. It was more or less of a freak contrivance. People spoke of it with a smile. Minstrels opened their performances by saying, "We come from Waterbury, the land of eternal spring"; and there is a story of a Waterbury owner in a sleeping-car, winding until his arm ached and then passing it to a total stranger, saying, "Here, you wind this for a while," with the result that the stranger placed a large order for Waterbury watches to be sold by his agency in China.

At the time that the Waterbury watch was well established, the world had advanced to a point fairly approximating the life of to-day. All the marvels of invention which had lifted so much of the earth's manual labor from the shoulders of mankind and which had been expected to shorten working-hours and to cheapen products until the standards of living of all classes would be raised through the possession of beneficial products inexpensively produced these had gone far toward establishing the factory system. Machinery had come into vogue in place of hand labor. The steam-engine, the sewing-machine, the railway, the steamboat, the cotton-gin, the threshing-machine and the harvester, were indispensable aids. Photography and typewriting were novelties no longer, and the phonograph was becoming familiar. Electricity had taken its place as one of man's most valuable servants, able to transmit his messages, furnish him with power, and turn his night into day. These are but a few of the countless improvements that had contributed to the rapid rise of this country as a manufacturing nation instead of one chiefly agricultural.

Millions had already found employment in the factories, the transportation systems, and other collective-labor establishments. Schools had multiplied throughout the country. Trains, for the most part, were run on schedule time. Business offices, accompanying the development of the great industrial concerns, employed thousands. The department store was beginning to appear. Public-utility organizations and government departments were growing complex and extensive.

Thus, in every direction a stirring impetus was being given toward those intricate modern conditions which depend upon the watch. The lives of nearly all people were beginning to be touched by affairs that demanded common punctuality a number of times every day the hour of opening factory, school, office or store, the keeping of appointments, the closing of banks and of mails, and the departure of trains. The times were bursting with need for a closer watch on time. From the industrial president to the common laborer and school-child the pressure of modern life, with its demand for punctuality, was making itself increasingly felt.

Yet, strangely enough, watches were still regarded as luxuries. It was not yet realized that they belonged among the implements which the daily life required of all. The notion still held that the watch was the mark of the aristocrat a piece of jewelry rather than an article of utility, a thing more for display than for use. And the prices of good watches, according to the standards of the day, were such as to perpetuate the idea.

It is no wonder then that, in spite of its crude characteristics, the low-priced Waterbury watch attained a considerable sale. A watch was a novelty, an uncommon possession among average people, and anything approximating a real watch was assured of a large sale if within reach of the ordinary purse. Therefore, the commercial failure of the Waterbury Watch Company involves something more than a mere business failure. Here is something which textbook economists may well undertake to explain, since the article was good, the need unsupplied, the competition feeble, and the profit satisfactory. The Waterbury watch enjoyed an initial success but, in spite of satisfactory quality, its sale gradually fell away, until, notwithstanding several refinancings and changes of management, undeserved failure ultimately overtook the first low-priced watch-venture. It was not the manufacturing problems, such as had overcome Howard and had sorely tried Dennison, but the problems of distribution which were the undoing of the Waterbury Company, and here the importance and power of the middleman stand out in an instructive way.

The conditions of the age demanded a cheap watch. Things to come could not eventuate except through the ability of everyone to measure his minutes. Almost from its first announcement, the Waterbury sprang into demand, but later succumbed to false policies of sales. Eagerness for the large and easy orders, which were momentarily attractive but finally fatal, spelled ruin.

When first put out, the watch was sold through stores at a very moderate price and proved to be such a sensation that it suggested itself to ingenious merchants as a trade-bringer when offered as a premium with other goods.

Sam Lloyd, the famous puzzle-man, was among those who saw this possibility and he devised a scheme which resulted in the giving-away of hundreds of thousands of Waterburys; it consisted of puzzles printed on cards. These puzzles were so simple and yet so cleverly designed that while anyone could solve them, each thought himself a genius for his success in doing so. Lloyd's idea was to take his puzzles to clothing stores all over the country and sell them with watches, in order that those dealers might distribute the puzzles all over town, together with an announcement of a guessing-contest. Each successful contestant, upon return of the puzzle with its solution, was privileged to buy a suit of clothes and get a Waterbury watch with it free of charge.

Such was the magic of a watch in those days that the Waterbury boomed the business of hundreds of clothiers, who, as in nearly all something-for-nothing schemes, were careful to add more than the cost of the watch to the price of the suit. Nevertheless the idea took so well that Lloyd spread it into Europe, China, and other parts of the world. Thus, the Waterbury watch became a familiar object in many lands. Adaptations of the scheme, applied to other wares, were carried out by him and by others until giveaway propositions became the main channel of distribution for these watches. For a time, such methods flourished and the regular trade of ordinary watch-dealers correspondingly languished. But, finally, the scheme-idea lost its novelty and pulling power. People would not forever buy clothes in order to get watches. In the process, the Waterbury name had become a byword for tricks in all trades. Shoddy clothes at all-wool prices had become associated with it in people's minds. They stopped buying these watches in ordinary stores because others "gave" them away. Regular dealers cut the prices to get rid of their stocks, and this led to further demoralization because customers never knew whether or not they were buying at the bottom price. Dealers could make no money on them under such market conditions and, because of this and of their shady association with give-away deals, the Waterbury name became a stench in the nostrils of the legitimate trade.

Thus, when the scheme-trade died away and the company again turned its attention to the watch-dealers whom it had forgotten in the flush of its easy success, it found no welcome. It had forsaken its source of steady customers and was now forsaken in return. After floundering about in several further reversals of trade policy and causing the loss of further investment for its backers, the Waterbury name was abandoned and the company reorganized as the New England Watch Company. As such it ventured into new fields of watch manufacture and offered an elaborate variety of small and fancy watches and cases, and numerous models, sizes, and styles of movements sold on vacillating marketing policies. Never did it attain a genuinely sound footing, however, for it vacated its field of fundamental and distinctive usefulness, viz., the production of a reliable, low-priced, simple watch, to meet the advancing requirements of its day; it had gone back to the view-point of the watch as an ostentatious or ornamental bit of vanity. Hence the old Waterbury business was compelled to close its doors, and in the fall of 1914, the first year of the Great War, was bought out at a receiver's sale by a firm who had replaced it in the field of supplying watches for the masses. This firm rededicated the organization to its original mission, modernized its mechanical equipment, and revived the Waterbury name after a lapse of twenty years, until to-day, through the employment of judicious sales-methods, the factory is more successful than ever it was in its earlier days.

THE WATCH THAT MADE THE DOLLAR FAMOUS

The next development is so typically American that it is difficult to picture it as occurring in any other country.

Heretofore, the history of timepieces had been that of an easily traceable evolution, for each of its steps had grown naturally out of those before it, and the various improvements had been made by mechanics trained in the craft. Yet now, strange to relate, two young men from a Michigan farm, with no mechanical training, entered the field almost in a casual manner, and in less than a generation not only became the world's largest manufacturers of watches but effected the most radical development in the whole story of telling time involving, as it did, the introduction of interchangeable parts, quantity-production, and a low price.

These results might seem at first, to be due to a matter of accidental good fortune. On the contrary, they were an example of evolution quite as logical as any that had preceded and were perhaps even more significant. The whole development came as the direct product of observation, analysis, initiative, perseverance, and hard work the element of good luck being conspicuously absent.

All history gives evidence of the occasional need of a new impulse derived from outside, and bringing with it a fresh view-point. There seems to be a tendency in human enterprise for any development after a time to lose its original rate of speed and to spend itself in complexities. The people who have brought it about appear to lose their power to see things simply and in a big way; and, on the contrary, they grow technical and occupy themselves with minor details. Whereupon the progress of development becomes slower and slower, and threatens to stop entirely. Then over and over again, there is the record of the advent of some fresh new force from an unexpected direction which restores youth and vigor.

In the last decade of the nineteenth century, watch-making seemed ready for such an impulse. As we have already seen, it had long been developing from within along technical and professional lines. Excellent and costly timepieces that were marvels of accurate mechanism had been produced. That part of its work had been well done, but the industry was in danger of losing its human touch. Watches were being viewed more as articles of manufacture and merchandise than as of wide-spread human service in meeting a general public need.

In a sense, therefore, the industry was unconsciously waiting the coming of a non-technical man who knew the public at first hand and understood people's requirements, who was not fettered by tradition, who had a vision of universal marketing and distribution, and who was not held back by a fore-knowledge of difficulties. It was exactly this vision which Robert H. Ingersoll had of the industry and he developed it with the assistance first of his brother, Charles H. and later of his nephew, William H. He did not "discover" the dollar watch, as many think, but grew toward it during the course of a dozen years.

It came about, as already stated, in a manner that was typically American. Young Ingersoll left his father's farm near Lansing, Michigan, in 1879, at the age of nineteen, and went to New York to seek his fortune. He was entirely without technical training save in farming, but he had a considerable first-hand knowledge of the needs and desires of what Lincoln called the "common people." Finding employment for a time, he saved One Hundred and Sixty Dollars, and, with this large capital, started in business for himself in the manufacture and sale of rubber stamps. Before long he was able to send back to Michigan for his younger brother, Charles H. Being of an inventive turn of mind, he devised a toy typewriter which attained a considerable sale as a dollar article. This was followed by a patented pencil, a dollar sewing-machine, a patent key-ring and other novelties of his own creation.

In the course of time, the products of other manufacturers were added to the list. Thus the brothers soon found themselves with an embryo manufacturing and wholesale jobbing business. The business grew, and the next development was that of a mail-order department. In this branch they were pioneers and preceded by some years the famous mail-order houses of Chicago and elsewhere. Their catalog ran into editions of millions of copies. Next, the Ingersolls became pioneers in another sales-plan. They developed the chain-stores idea, starting with a retail specialty store in New York, and following it with six others. Incidentally, they found themselves among the largest wholesale and retail dealers in the country in bicycles and bicycle supplies.

All of this was a strange but none the less effective preparation for watch-making and the marketing of watches by millions. Robert Ingersoll, who had remained in the selling and promoting end of the business, knew little about watches, but since he was constantly engaged in traveling about the country and in talking with merchants and others, he was gaining a great fund of knowledge as to human needs and market possibilities.

Presently he became convinced that his business, in spite of its prosperity, lacked something vital. He grew dissatisfied with handling a succession of unimportant novelties. It began to dawn upon his mind that these things were hardly worth while as a subject for a business, since they satisfied only passing fancies on the part of the public. He must find something which was really worth while, something which filled a real human need on a large scale and yet in a new way. If this something could be found, and the incredibly large buying power of the great American public could be focused upon it, there was hardly any limit to the business which would result.

When this belief had crystallized in the form of a definite conclusion, he began at once to search for the "big idea." The "big idea" had long been waiting for him to reach this state of mind. It had been looking him in the face for many days had he but been ready to perceive it.

On the wall of his room in a Brooklyn boarding-house there hung a very small "Bee" clock. It was unobtrusive and apparently unimportant. He had glanced at it hundreds of times with no thought beyond that of learning the time. Suddenly, it ceased to be a clock and became an open door into the future. Its ticking became articulate with a new meaning.

This picture shows one corner of the huge plants which produce twenty thousand Ingersoll watches a day.

"Everyone wishes to tell time," it said. "There is not one of the millions who crowd the cities, travel the highways, or spread over the country districts, who does not wish repeatedly during his waking-hours to know what time it is. Sometimes he is in sight of a clock, but more often he is not. Here and there is a man with a watch in his pocket. That man has a chance to be efficient; but good watches cost money, and most people cannot afford them. Here am I, a tiny little ticking clock; I am a good timekeeper and I am cheap. Make me a little smaller, sell me for a dollar, and you can put the time into everyone's pocket."

At this point, the non-technical man, who knew nothing about watches, but who understood human needs, realized that something had happened; he pondered deeply and began to investigate. He took the little clock to a machinist in Ann Street, New York, and together they studied the possibility of reducing it in thickness and diameter. Presently it was discovered that both the New Haven and the Waterbury Clock Companies had already produced articles that embodied these conditions. This somewhat checked enthusiasm until it was recalled that neither of these products was an especial factor in the time-telling field. The manufacturers had merely made mechanisms; they had not grasped the Big Idea of universal service.

The timepiece of the Waterbury Company was the smaller, and Robert Ingersoll decided to test his mail-order market, buying first, one thousand clock-watches at eighty-five cents each, and afterward contracting for ten thousand more. These articles were offered in the mail-order catalog for 1892 at a dollar each, for the sake of price-uniformity with the other dollar specialties upon which the firm was concentrating. This was done, however, in a small way. It was not desired to sell too many on such an unprofitable margin, but merely to test the dollar-watch idea, hoping that manufacturing charges might ultimately be brought down through quantity production.

These so-called "watches" must not be confused with the Waterbury watch; that, as already described, had been the output of another company. The "watches" marketed by the Ingersolls and bearing their name were in reality thick, noisy, sturdy little pocket-clocks, wound from the back. They were crude and clumsy affairs compared with present-day styles but were, nevertheless, reliable timekeepers.

The public responded to the idea of dollar watches, although these proved to sell faster in gilt cases than in nickel, and still faster when a five-cent gilt chain was added. The next year, came the World's Fair in Chicago and the odd little mechanism with an appropriate design stamped upon its cover attracted some attention from the visitors.

Thus was born the Ingersoll watch, although it bore slight resemblance to the watch of to-day. This is due to the fact that an immediate policy of experiment and improvement was inaugurated. During these changes, however, several points remained fixed. One of these was that the watch must be in no respect a plaything, but a practical accurate timekeeper, not liable easily to get out of order. The second was the definite association with the price of one dollar, so that it became possible to refer to it humorously as "the watch that made the dollar famous;" and the third was that it should have a sturdy ruggedness of construction that would defy ordinary hard usage.

Each of these points had its social value that of the last-named being the fact that the dollar price put the possession of a real timepiece within the reach of multitudes who were engaged in forms of activity wherein a delicate timepiece would be apt to get out of order.

The Ingersolls soon became convinced that they had a worthy object for promotion, and they did not entertain the slightest doubt as to the existence of a waiting public. There passed before their minds a picture of the millions of farm-boys who did not know when it was time to come into dinner, of the millions of working-men who had nothing to guide them in reaching the factory on time, of millions of clerks and school-children and of still other millions comprising the bulk of American homes where more good timepieces were needed.

Their problem, therefore, resolved itself into two main divisions those of manufacture and those of sale. The manufacturing end involved a contract with the great plant of the Waterbury Clock Company, by which this factory was to produce the goods according to the specifications and under the name, trade-mark, and patents of the Ingersolls. This arrangement continues to this day, but has been supplemented, as the line has become more extended, by the acquirement of two factories of their own, one in Waterbury, Connecticut, and one in Trenton, New Jersey. To-day the three plants produce an aggregate of about twenty thousand watches a day. Before such manufacturing results could be obtained, however, there were many structural problems to be solved. It was not so easy as it sounds to build a practical and accurate watch within the narrow limits of a dollar and still leave a profit for both the manufacturer and dealer.

The solution began with the adoption of the "lantern-pinion," but the principal difficulty was that which had baffled both Howard and Dennison the problem of producing the extremely minute separate watch-parts in large quantities by machinery, and yet with such exquisite precision that all parts of one kind should be absolutely interchangeable. By dint of unwearied patience and much scientific research, this problem was finally solved, and it is said that Henry Ford got his idea of quantity-production from the manufacture of the Ingersoll watch. Incidentally, it was demonstrated that low production-costs carry with them high wages. In the field of watchmaking, no element was more necessary than the skill of well-paid workers.

In the meantime, the public was waiting, but it did not know that it was waiting. It was going about its business quite unaware that mechanical and manufacturing problems were being solved in its behalf. There were no eager millions standing about demanding watches in order that their lives might be run more closely upon an efficient schedule. Therefore, simultaneously with the consideration of mechanical and manufacturing problems came those of sale, which will be discussed in the next chapter.

PUTTING FIFTY MILLION WATCHES INTO SERVICE

If this were purely a story of the development of timepieces as mechanisms, there would be little to add to the preceding chapter, save to detail the refinements and improvements by which a cheap, clumsy, but reliable watch gradually discarded its defects, while retaining its virtues, and the manner in which it developed into a variety of styles and sizes. Essentially, however, this is a story of Man and Time, of human needs as served by timepieces. The most perfect piece of mechanism in a showcase is like a stove without a fire; it is a mere possibility of service, whose value does not begin until it is set to work.

We have arrived, then, at a time when a small percentage of the total population carried accurate timepieces and was able to profit by the more efficient adjustment of its actions thus secured. We have seen how the promising experiment of the Waterbury Watch Company failed in an attempt to equip the masses with watches, principally through defects in its system of distribution, and we have noted the appearance of another low-priced watch dedicated to a similar experiment.

It is obvious, therefore, that if the Ingersoll firm has already been able to place fifty million separate watches in the service of humanity, something unprecedented must have taken place in the all-important field of distribution. It is significant that Robert H. Ingersoll first called his watch the "Universal;" indeed, his chief contribution to the development of the watch is the idea of universality, a word that makes us think more of people than of manufacturers' methods. Having, then, a watch that was universal in its possibilities as well as in name, and being keenly aware, through his own tastes and experiences, of the needs of the vast mass of the public, his greatest problem became that of universal distribution; in short, it was a selling-problem. At first, there could be no definitely formulated plan; various methods must first be tried out. From these experiences there gradually arose an adequate system of reaching the millions of people who needed watches.

In this, Mr. Ingersoll had effective cooperation. He was the pioneer, the salesman, the promoter, the one who knew men in the widest sense and had the faculty of getting results. His brother, Charles H., was the internal administrator and constant counselor. Later, there was added to the firm a nephew, William H., who was both a student and an analyst. He scrutinized trade-tendencies, deduced theories from what he saw, and gave them wide application in actual tests. Together the members of the firm worked out sales-principles of equal opportunity and equal treatment words that had long constituted a slogan in politics but were something of a novelty as applied to business. In other words, they based their plans upon the consumer rather than upon the factory, and upon the idea of goods sold through the trade rather than to the trade. It took some time, however, to perfect their system of distribution but, when finally developed, it was the outgrowth of wide and varied experience.

The firm made its first sales-efforts on the watch through its own mail-order catalog. The results brought some encouragement, but proved that in itself this method could never bring the volume of sales necessary for a high-geared, uniform quantity of production.

The next recourse was to the so-called "regular trade-channels" the jobbers and retailers. But these dealers displayed little interest. They were not promoters of new lines, but distributors of those for which a market already existed. The jobber sold what the retailers required; the retailers what the public demanded. Robert Ingersoll's original loud-ticking watch impressed them more in the light of a curiosity than as a trade-possibility. In particular it failed to appeal to the jewelers, since they felt it to be out of keeping with the beauty and value which characterized their stocks of jewelry and silverware. They reasoned, also, that sales of the new timepiece would interfere with those of their higher-priced watches, thus failing to grasp the fact, since proved to be true, that its use would greatly enlarge the sphere of their sales through cultivating a general watch-carrying habit.

Here is represented the final stage in the development of modern timepieces. Though of graceful lines, they are designed for accuracy and utility, and are ranged in price to fit every purse.

Some effort was made with outside trades, but these generally considered watches to be out of their line. Nevertheless, in the course of time, persistent effort began to bring results. Occasionally jobbers made purchases, and here and there a jeweler or hardware dealer offered the watches for sale. When the firm felt justified in spending some money for advertising, the public began to learn at first hand of the Ingersoll watch, and the sales gradually increased. Many people, however, expressed doubt as to the quality of a timepiece that could be sold for a dollar, and the Ingersolls replied with a guarantee that has since become famous.

Then, in the natural course of business, competition developed from the marketing of inferior goods, and the firm found it necessary to place its name on the dial for purposes of identification. In spite of all difficulties, there grew up in course of time a very considerable public demand. Whereupon certain dealers undertook privately to raise the price in order to increase their profits. This situation was met by emphasizing the price more prominently on the boxes and in the advertising, a policy which soon put an end to price-raising but led, in some instances, to the even greater difficulty of price-cutting. The better known became the price, the greater became the temptation to dealers of a certain class to advertise its reduction in order to bolster up "bargains" upon other goods. This naturally demoralized the sales of neighboring dealers and caused them to lose interest in the line. Thus, instead of increasing the sales, the reduced price proved a serious selling obstacle.

The same difficulty has been encountered by other manufacturers of widely advertised goods, and some of them have sought through the courts to compel adherence to their prices, the argument being, as in the case of the Ingersoll watch, that price-cutting does not serve the interests of the public but tends to interfere with sales since it obstructs the channels of distribution. At this writing, the question in its legal phase has not yet reached a final decision in the courts, but the Ingersolls have solved it in a practical way, since their trade-policies have brought about the voluntary cooperation of the retailers.

Such cooperation, however, was not to be attained at once. It came about through much study and after much experience. It involved the assembling of a large amount of data upon commercial economics and a deep inquiry into the fundamental principles of retail distribution. It proved necessary to weigh and compare recent and important factors in the retail situation. For example, because of the fact that so many manufacturers were giving indiscriminate discounts for quantity purchases, it had become profitable to establish huge department stores, chain-stores, and mail-order houses whose scale of operation made it possible to handle goods in large amounts. For a time, the Ingersolls, in common with other manufacturers, gave discounts for purchases in quantity; later, as the business grew and its distribution problems were more scientifically studied, they saw more clearly the way in which the principles of equal opportunity and equal treatment could be applied.

It was in this spirit that the firm began to ask itself whether the large distributors were really more efficient than the small retailers; whether they actually earned the extra amount which they were paid for selling each watch, and whether it would be a healthful thing for the country if all retail business were transacted through such organizations in short, whether restrictions to such a system were really consistent with the theory of commercial democracy.

Approached from this standpoint, the answer was found to be in the negative. A careful research among stores in all sections of the country showed unmistakably that the cost of selling in a small store was actually less than in the department store, the chain-store, or the mail-order house. Viewing the sale of each watch as an individual transaction, it was seen that a small store in some far-off country village gave quite as valuable service as did a large store in a metropolis, and therefore should be paid as much. Consequently, the Ingersolls introduced a selling-plan which, under the conditions, was as revolutionary in the field of retail distribution as the discovery of Galileo had been in that of clock mechanism. Yet it was merely that of a flat-price schedule; in other words, it was a provision that the dealer buying one dozen watches, or even one single watch, should pay exactly the same price as the dealer who bought ten thousand. Quantity discounts were definitely abandoned.

Naturally, this plan met with cordial response from the countless small retailers scattered throughout the length and breadth of the country, and the close relationship thus established led to other logical developments in the way of cooperation, such as that of display devices suited to the needs of these dealers, a simplified accounting system to increase their efficiency, and various measures of a similar nature.

In the meantime, a constantly increasing advertising appeal resulted in a rapidly growing demand from the public, and this, in turn, made possible the assuring a uniform quantity of output, which was in itself the basis necessary for maintaining uniform quality. Thus practical experience and scientific trade-study were formulated into what has come to be recognized as a definite commercial philosophy, namely, that of uniform quality, uniform quantity, uniform demand, uniform price to the dealers and uniform price to the consumer a statement of principles in which, as in the works of a watch, each part must be geared to every other to insure effective operation.

During the time that these business principles were being formulated, the line of watches was also in process of development with the goal of universality in view. Thus, it was presently realized that while the dollar watch was essentially a man's timepiece, watches were also needed by women and by children. Accordingly, smaller models were developed to meet these needs. At a later date, the Ingersoll business principles were extended into the field of jeweled watches, when the factories of the Trenton Watch Company and the New England Watch Company were acquired. At the date of the present writing, there are more than a dozen models, each of which is adapted to a different need and use, but the manufacture of no model is undertaken unless there is a market for at least a thousand watches a day.

And the latest development as this is written is the time-in-the-dark watch.

Do you recall a soldier in the "foreword" waiting in the darkness for the perilous moment to go "over the top" with his eyes fixed upon the luminous hands and figures of the watch strapped to his wrist? This watch may now be named; it was the "Radiolite." How it came into existence in time to go into the Great War is a story in itself.

This story is the latest step in that steady progress of democratization by which accurate timetelling, once a privilege of the few, became the possession of the many.

A good many people wish to tell time in the darkness as well as in the light, and if these people could afford to, they bought expensive repeaters. Such watches, however, cost hundreds of dollars, so that while telling time in the light had come within the reach of everyone, telling time in the darkness was still possible for very few. Therefore, the watch could not yet be held to be of equal service to all humanity in every one of the twenty-four hours. This equal service at any moment was finally made possible in a somewhat extraordinary manner.

In the year 1896, Monsieur and Madame Curie startled the world with the discovery of radium. They found that certain substances emitted rays that would pass through solid matter as light passes through glass or as the wind blows through a screen. They were finally able to secure tiny quantities of a whitish powder, salt of radium, which gave forth an energy that acted upon everything brought near to it and this energy they calculated, would be protected uninterruptedly for three thousand years. Up to the present time, radium and radioactivity are subjects of constant study and research, but radium exists in such small quantities and is so enormously costly that comparatively few have had a chance to experiment with it.

It seems a little strange to think of using the most precious substance in the world many times more costly than diamonds in order to bring time-telling-in-the-dark within the reach of every person, but this is exactly what has been done.

People had long been experimenting with paint made from phosphorous in order to give off a glow in the darkness which would be sufficient for time reading, but phosphorus has its limitations; it must first be exposed to light before it is taken into the darkness, and if a watch-dial treated with phosphorus is buried in the pocket it cannot absorb enough light in the daytime to be luminous at night. With radium, however, the problem was solved. It was found that this amazing substance would affect certain other substances, causing them to shine for years in the darkness by means of their own light.

Thus it became possible to develop a luminous coating which the Ingersolls applied to the hands and figures of their "Radiolite" watch and, presto! the problem of telling time in complete darkness was mastered to the advantage of every buyer. The inexpensive watch revealing the hour with equal visibility in inky darkness as in bright daylight had become a reality. In passing, it is interesting to note that the experiments with the watch-face led to many other developments, such as luminous compasses, gun-sights, airplane guides, and the like.

Then came the World War, and the wrist-watch which had been often ridiculed as effeminate (although it is hard to explain why, since it was first adopted as an obvious convenience in the Army and on the hunting-field two of the most masculine spheres of activity it would be possible to imagine) was seen at once to be the most easy means of knowing the time in actual warfare. Millions of watches, consequently, were strapped to wrists of soldiers and sailors, and the obvious advantages of the luminous dial placed it in enormous demand. Thus it came about that the scene described in the opening pages was typical of countless instances upon various fronts.

Although a matter of surprisingly few years, considered chronologically, there is a long distance, measured by the scale of progress, between the moment when a young man, glancing casually at the clock on his bedroom wall read wonderful possibilities in its face, and the time when the firm he founded was able to take note of such achievements as these:

Many a soldier waited in the darkness for the perilous moment to go "over the top," with his eyes fixed upon the luminous hands and figures of his Ingersoll Radiolite.

Factory facilities producing an average of twenty thousand accurate watches a day; distribution facilities including the cooperation of a voluntary "chain-store system" of more than one hundred thousand independent retailers, all operating upon a common plan and under common prices; a product that has come into the most wide-spread use not only throughout the United States but in the farthest regions of the inhabited earth which has, in fact, in itself served to turn back the tide by which watches formerly flowed from Europe into America, so that it now proceeds from our shores toward those of Europe and other lands; a name which has become as well known as any in commercial and industrial life, and better than all, the appreciable raising of the efficiency of the human race through universally promoting the watch-carrying habit and putting fifty million timepieces into service. It is altogether an Aladdin tale of modern business.

THE END OF THE JOURNEY

Did you ever, at the end of a journey perhaps across water, or up to the top of some high hill look backward to the place from whence you came, and wonder that it seemed so far away?

Now as we have completed our journey together through the history of man's struggle to gain knowledge and control over time, we are impressed with the great contrast between Time as it was to mankind in the beginning, and Time as it is to us to-day.

The caveman, with whom we began this story, lived close to nature, taking his sense of time from her as he took all else. Morning was when the light came, and he waked and was hungry; noon was when the sun was highest, and night was the time of lengthened shadows and the state of darkness. We see these same things, but, for us, they have not the same meanings. We count the time by hours and minutes, and we reckon these by machines which we have made, called clocks and watches. These mean so much more to us that, when we set all the clocks forward another hour to save daylight, it seemed to us as if we had changed the actual time. It was practically as if we had performed the miracle of Joshua, who in Bible story, made the sun stand still, or the miracle of Isaiah, who made the shadow go back ten steps on the dial of Ahaz. After a few days, we did not feel as if we had set the clocks; we felt as if we had made the sun wait for us, and the very day come earlier.

And so it is with the seasons. The caveman called it spring when the swallows came, and autumn when the leaves changed their color. But we judge of these things by the calendar; we say that the spring "is very late this year," or that the "leaves are beginning to turn early." We have a proverb that one swallow does not make a summer; no, nor do all the swallows, so far as we moderns are concerned. It is summer for us upon a certain day, no matter what the swallows do, but for the caveman, summer was when the swallows came, whenever that might be.

It is like that to-day among primitive peoples. The Turk who listens for the crowing of a cock or the braying of an ass to tell him of the hour, or calls the cat to him to look at its eyes and judge the time by the shape of their pupils he is more like the caveman in this than like ourselves. So is the South Sea Islander, who knows the season of the year from the direction of the trade-winds. So is the patient savage, who cares little as to how long he must wait for the creature he is hunting to come near the spot where he lies hidden.

How different it all is with ourselves! We rise at a certain hour, and so many minutes later we have our breakfast. At such a time, we must be at work. Our work itself is all made of appointments one after another, or of tasks to be finished within a certain time. Our meals, our hours of rest, our meetings with our friends, our recreations, and our pleasures all these, until, again, at a certain time we go to bed, in order that so many hours of sleep may make us fit for the next day, are measured by the clock and counted out by the tick of a toothed wheel or the regular swing of a pendulum.

We say that the savage has no sense of the value of time. We have, and it is by that fact largely that we are better off than he. Value means measure; you cannot value a thing unless you can measure it exactly. And so because we can measure time, we can see what time is worth to us, and make it worth more. The savage keeps an appointment when he happens to make one. But we, because we know how long it takes to reach a certain place, or how long a time we need or wish to spend with a certain man, can make and keep many appointments. We can travel like the wind from place to place, because in measuring time we can measure speed, and therefore we can make speed safe and possible. We can talk to a friend a thousand miles away, or signal by electric waves around the world. We do these things because our sense of time has told us that the old way of sending letters and messages was too slow. And so we have set to work to invent ways that should be quicker. We should never have had the telephone, the cable, or the wireless, unless we had cared about time and been able to measure it.

The caveman lived, perhaps, as many years as we but how much did he do in those years? We, who have learned to measure years and to allot each day or hour to sundry tasks, have made ourselves able to do far more in a life-time many times more. We do not live a greater number of years, but it is as if we lived many lives in one. We speak of time as we speak of money, of saving and wasting and spending. Well, Time is Money, as Ben Franklin said, but it is something more Time is Life. And we think of our lives as so much time at our command, and therefore we can make the most of them. The gulf between us and the primitive men is a contrast of living less or more, and our more life comes in great measure from our having learned to measure time.

Everyone has read the story of Aladdin and his wonderful lamp. You will remember that the poor boy came into possession of a lamp which quickly made him the richest and most powerful person in the world, since, through owning it, he could control the service of a mighty genie, able to perform the most incredible tasks.

The modern man every man is something like Aladdin, only he is much more powerful. He has the genie of steam to work for him when he pulls the lever, and the genie of electricity ready to serve him if he but press a button. He has many other mighty servants that modern science has given to him, but greatest of all, most useful of all, is the Slave of the Watch which lies in his pocket mighty Time himself.

This ability to record time and therefore, to control it, is perhaps the greatest of all man's triumphs. Only see what it has done for him! Have you ever thought of yourself as a person of no special importance? why, you have far more actual power than was possessed by Alexander the Great, Julius Caesar, or Charlemagne!

You can command forces and can accomplish results that would have made any of these proud autocrats stare in wonder. If you do not stand out above your age, as they did above their ages, it is simply because millions of other people besides yourself also possess these powers. It is undoubtedly true that we are to-day a race of giants, and it is also true that each of our powers is directly or indirectly due to the common fact that we all can keep track of time. For consider that what mankind can accomplish to-day depends upon the ability of people to work together, and that working together would cease if people had no accurate means for telling time.

For example, you make a railway journey upon a matter of importance to you. The first thing that you do is to examine a time-table on which is shown the minute when the train is due to leave. You calculate to yourself how many minutes you must allow for reaching the station, and then look at your watch to see how long you will still have for other work. If you had not watch or clock, or you were dependent merely upon the position of the sun, you might go to the station several hours ahead of time in order to be "on the safe side." During the hours thus saved you can accomplish a great deal of work. It is as though your day had been made several hours longer.

Unseen in your pocket, your watch ticks steadily. You trust it absolutely, and you know that it will be faithful to its trust. Occasionally you glance at it and, when the hand reached the limit of safety, you start for the train. You reach the station three or four minutes before train-time and find the tracks clear; no train is in sight.

This however, does not cause you the least uneasiness. You merely take your watch from your pocket and look expectantly up the line. Perhaps a minute before the train is due, you hear a distant whistle, then the approaching roar of wheels upon the rails, and, just as the watch-hand reaches the proper moment, the train itself whirls round the curve and draws up to the station, exactly on time.

As you proceed upon your way, you notice how other people at other stations are also meeting their schedules and conserving their time. You see the conductor glance at his watch as he gives the engineer the starting-signal. You realize that the whole transportation system is merely an enormous piece of clockwork and that it, in turn, is a part of the vaster clockwork of modern civilization.

Turn where you will, there is nothing that you can do and nothing that you can use which is not dependent upon the ticking of clockwork. The locomotive which pulls your train, the cars in which you ride, the rails over which you pass, all of these are products of factories, but the factories are run upon the time-basis; there is no other way in which they could be run.

The workmen in these factories leave their records upon time-clocks when they come and when they go. If the workmen were not there at the same time, the work could not be done, since most of modern work depends upon the ability of people to work together at the same task. Even if one man were late, it might lose time for many. The clothes that you wear come from other factories where other workmen have time-clocks and watches. The buildings that you see from the windows were put up on the time-basis and were paid for according to the movement of the hands upon watch dials.

Without the ability to record time, and, therefore, to control it, the complex web of human activity would become hopelessly tangled.

You buy a newspaper, making sure that you are getting the latest edition, and it is at once as though you looked into a great mirror reflecting the activities of all the world, but all of the dispatches bear a date-line, and many of them are also marked with the hour.

Before the days of newspapers, people felt themselves to be a part of the lives of their own immediate neighborhood and knew only vaguely of what went on at a distance, but now each day one feels himself to be a part of the great human family and can sometimes make his plans with reference to things that may be occurring thousands of miles away. But the newspaper itself is a product of clockwork; there is perhaps no institution whose workers keep closer track of the passage of the minutes.

In view of all these things, does it seem too much to claim that if all the timepieces in existence were destroyed and men were given no other means for telling time, civilization would swiftly drop to pieces and man would find himself traveling backward to the conditions of the caveman?

But there is one thing in our modern timekeeping which we still have in common with the first men who ever kept the time. We still go by the sun and the stars and refer all our measure to that apparent revolution of the heavens which we know to be really the motion of our world itself. As did those wise men of old Babylon, so do we even now, spying upon the mighty master clock of the universe to correct all our little timepieces thereby. A man sits alone in an observatory, with his eye to a telescope. That telescope is of a certain kind, called a "transit." It is fixed upon the meridian, the north-and-south line in the sky over that place. And a thread of spider-web across the lens marks for him the exact position of the line, in the very middle of his field of view. So as he watches, he can see one star after another come into view at one side of the glass and pass across it to the other side and disappear. He is watching the world go round.

A certain star appears, one which his calculations have told him will cross the meridian at a certain particular instant. Beside him is an electrical device connected with a clock, which marks off seconds at intervals round a revolving drum. The star draws nearer to the center of his field. As it crosses the hair-line, the observer touches a key, and the precise instant of its crossing is recorded upon the drum, to within a fraction of a second. Since the clock has marked its record of the seconds there, the clock can be corrected by the star.

Now, if that man had been a priest in Babylon, he would have kept his knowledge as a means of power to himself and to his equals. If he had been a dweller in a somewhat later age, he would have kept it to himself no less, either because people would not believe, or because the claim of too deep knowledge of the secrets of nature might put his life in danger. But he is a modern, and so his knowledge is for all who seek it.

On some tall building in a distant city, a time-ball hangs suspended at the top of its pole, and people pause to look up at it. They hold their watches in their hands. Upon the tick of noon, an impulse will come from the observatory, and the ball will drop. Then those who have been looking will set the hands of their watches and pass on. At the same instant, the news of noon will be flashed by telegraph across the land, and by wireless to ships at sea. The whole Western Union system will suspend business for a little, while the lines are connected and the observatory at Washington ticks off the seconds. Everywhere there are electric clocks, automatically controlled by some master clock, which, in its turn is governed by the observatory time. So we all, as a matter of course and without thinking, set our watches by the star. Civilization every day catches step with the heavenly bodies.

Back of all that we see of life, therefore, stands the great fact of measuring time, and those who are engaged in giving to man the instruments for this purpose have a special responsibility. Perhaps the ancient peoples were not so far wrong when they permitted time-telling to be a privilege of the priests. It is far more than a matter of moneymaking; it is a fixing for humanity of the standards of daily life; it is a duty which lies at the foundation of modern efficiency; it is even a sacred trust.

Therefore, the man who makes or sells unreliable timepieces is false to his trust. Through his action people are thrown out of adjustment with the world about them, and they, in turn may seriously interfere with the plans of many others. It is hard to believe that there are some people who still look upon a watch as "jewelry," or that there are some dealers who are more interested in the watch-case than in the movement it contains.

The watchman of olden times was a public officer. He was chosen for his reliability, and people felt confidence when he called the hours. The watch-dealer of to-day is in a somewhat similar position; he has a serious duty to his community. He is not chosen by the public, and yet, even more than the watchman, he is a public servant since the watches that he puts into people's pockets are their principal means of adjustment to the busy affairs of life. In a sense, he supplies them with the basis of their efficiency. His duty is that of supplying the largest practicable degree of accuracy to the largest possible number of people. The Slave of the Watch will not obey the owner of an inaccurate timepiece.

Time itself is elemental; it had no beginning, it can have no ending. It is like a great ocean which flows round all of the earth, and neither begins nor ends in any one place. But time for any man is exactly according to his use of it. It is as though a man were to go to the shore of the boundless ocean, with a tin cup in his hand. If he could get no more than a cupful of water, it would not be because of any limit in the amount available, but merely in his means for carrying it away. Should he have a pail, a barrel, or any larger receptacle, then the water would belong to him in a correspondingly larger amount.

Thus, time each day presents itself equally to everyone upon the earth, but some receive it in cups, some in pails, and some in barrels. Some make of their day a thing of no results, while others fill it with real achievement. Those who achieve are they who have learned to value time, and to make it serve them as the mighty genie that it is.

These are the wonders which Kipling had in mind when he wrote:

If you can fill each unforgiving minute
With sixty seconds worth of distance run,
Yours is the earth and everything that's on it,
And, what is more, you'll be a man, my son!

APPENDIX A

How It Works

Having traced out the history of the clock and watch mechanism all the way from De Vick's first clock and the clumsy old Nuremberg Egg down to the perfect time-keeping device which we have today, it may be interesting to look a little more closely at the result of so many years and so many inventions to see what its parts are, and how they are put together, and to observe how the wonderful little machine does its work.

Modern clocks and watches are nearly enough alike in their structure and way of working, so that if we understand the one, we shall easily understand the other also. The differences between them are few and slight and easy to explain. So let us take for our example a typical modern watch movement, which is easily the more beautiful and interesting mechanism of the two.

First of all, as we saw in the days of De Vick and Henlein, a watch, or a clock, is a machine for keeping time. So it must have three essential parts: first, the power to make it go; second, the regulator to make it keep time; and third, the hands and face to show plainly the time it keeps. Each of these three parts is itself made up of several others.

The power or energy which runs the watch is put in to it by the winding which coils up the mainspring. The outer end of this spring is attached to the rim of the main wheel (1) and after the spring is wound this wheel would whirl round and let the spring run down instantly if there was nothing to stop it. The teeth on this wheel, however, are geared into the second or center pinion (as shown in illustration at "A") which makes it run the entire movement while running down slowly instead of flying round and uncoiling at once.

As we will see later, the spring-power is transmitted through the train of wheels and the lever (7) to the balance wheel (8) which lets the escape wheel (5) turn a little each time it swings, while it simultaneously receives, by means of the lever from the escape wheel, the "impulse" or power which keeps it running. Thus the swinging of the balance lets the mainspring down gradually while drawing its power from it. The spring is made as thin as it can be and still have power enough to make the watch go. For a modern watch, this is about one flea-power. One horse power, which is only a small fraction of the power of the average automobile, would be enough to drive all the millions of watches in the world.

The center pinion into which the mainspring is geared is attached to its staff to which is also fastened the large center-wheel (2) so that the spring cannot turn this pinion without also turning the center wheel. But the center wheel is, itself, geared into the third pinion, which is attached to the third wheel (3), and this again is geared into the fourth pinion attached to the fourth wheel (4). The fourth wheel gears into the escape pinion which revolves with the escape wheel (5), so that none of these wheels or pinions can turn except when the escape wheel does. But there is a constant pressure from the spring on all of these wheels, which together constitute what is called the train.

The escape wheel, therefore, wants to turn continually and if it was not restrained it would revolve rapidly, letting the movement run down. But it is retarded and can only turn from one tooth to the next, each time the balance (8) turns. This action is secured by connecting the balance and the escape-wheel by means of the lever (7), one end of which forms an anchor shaped like a rocking-beam, called the pallet (6). In the pallet are two jewelled projections called the pallet-jewels which intercept the escape-wheel by being thrust between its teeth, letting it turn a distance of only one tooth at each swing of the balance as the pallet rocks back and forth.

The other end of the lever is fork-shaped, having two prongs. On the staff with the balance instead of a pinion as all the other wheels have, is a plain, toothless disc called the roller, from the lower side of which projects a pin or rod made of garnet. This is called the jewel-pin or the roller-jewel. The roller being fastened to the balance-staff, of course, turns just as the balance turns and with it the jewel-pin. And the lever is just long enough and is so placed that every time the balance turns, the jewel-pin fits into the slot between the prongs of the lever-fork carrying it first one way, and then, as the balance comes back, the other way. Thus the lever is kept oscillating back and forth, rocking the pallet and withdrawing one pallet-jewel, releasing the escape-wheel just long enough to let it run to its next tooth before the other pallet-jewel is thrust in to stop it. It is a beautiful thing, to watch, like the beating of a tiny heart, or the breathing of a small quick creature. The hairspring (9) almost seems to be alive. And indeed, it is in a way, the very pulse of the machine.

A Modern Watch Movement

A Modern Watch Movement

There is only one more important point to understand. You know how the power gets as far as the escape wheel from the mainspring, and how the motion of the balance lets the escape-wheel revolve a tooth at a time, but you have still to learn how the power which keeps the balance rotating reaches it from the escape-wheel through the lever. Here is the most interesting feature of a watch movement.

After the balance has been started, its momentum at each turn starts the lever when the jewel-pin strikes it, but unless the balance was constantly supplied with new power it would soon stop, and the watch would not run. It will be noticed, however, from the illustration, that the teeth of the escape-wheel are peculiar in shape and very different from those of the other wheels. The ends of the pallet-jewels are also cut at a peculiar angle.

Now, each time just before the jewel-pin starts to shift the lever from one side to the other, the latter is in such a position that one of the pallet-jewels is thrust in so that its side is against that of one of the teeth of the escape-wheel, keeping it from turning. But the instant the lever commences to move it begins to draw this pallet-jewel outward from the tooth until the corner of the jewel passes the corner of the tooth. Then the escape-wheel is released and the power that is behind it makes it turn quickly, and on account of the shape of the tooth, it gives the pallet-jewel a sharp push outward, swinging the lever, causing it at the other end to impart a quick thrust to the jewel-pin, thereby accelerating the speed of the balance and renewing its momentum.

Thus the balance receives the power to keep it in motion, swinging it as far as the hairspring allows. The hairspring then reverses it and swings it until the jewel-pin again starts the lever in the other direction, releasing the escape-wheel from which it receives another "impulse" and so on as long as the mainspring is kept wound. A watch in perfect time ticks five times to the second. That means 18,000 swings of the balance every hour, or 432,000 in a day. And in that time, the rim of the balance travels about ten miles.

A clock is essentially only a larger and stronger watch, just as a watch is a clock made small enough and light enough to be carried about conveniently. But the working of the two is practically the same. They are but different members of the same family, varying types of one time-keeping machine which is among the most ingenious and valuable things that man has made.

One interesting thing to know about a watch is that if it is keeping good time, it will serve for a fairly accurate compass. So if you are ever lost in the woods, your watch may help you out again. Lay it flat face upward, and point the hour hand toward the sun. Then South will be in the direction half way between the hour hand and the figure 12, counting forward as the hands turn in the morning hours, and backward in the afternoon. This is because the hour hand moves around the dial just twice as fast as the sun moves around the sky, making a full circle in twelve hours while the sun makes its half circle from horizon to horizon.

Now, the sun is always to the southward of you as you are anywhere north of the equator. At noon, the sun is practically due South. At that hour, both hands of your watch are together on the figure 12 and the hour hand pointing at the sun points in that direction. At 6 a.m. the sun is nearly East, so if the hour hand, now on the figure 6 is pointed eastward toward the sun, then South would be in a line just over the figure 9. At 6 p.m., the sun being in the west and the hour hand pointed at it, South would be half-way back toward the figure 12, or just over the figure 3. For other morning or afternoon hours, the same reasoning holds true.

APPENDIX B

Bibliography
Adjusting, Practical Course in Theo. Gribi. Jewelers' Circular Publishing Company, New York City, 1901.
American Clockmaking Its Early History Henry Terry. J. Giles and Son, Waterbury, Connecticut, 1870.
American Watchmaker and Jeweler, The (An encyclopedia.) H. G. Abbott. Geo. K. Hazlitt and Co., Chicago, Illinois, 1891.
American Watchmaker and Jeweler J. Parish Stelle. Jesse Haney and Co., New York City, 1868. Revised Edition, 1873.
Ancient and Modern Timekeepers Reprint from Harper's Magazine, July, 1869. Albert D. Richardson.
Annuaire Suisse (de l'horlogerie et de la bijouterie) Supplement gratuit de l'Annuaire du commerce suisse. Geneva, Switzerland, 1912.
Artificial Clockmaker, The (Fourth edition with large emendations.) Wm. Derham. James, John and Paul Knapton, London, England, 1734.
Ausführliche Geschichte der Theoretisch-Praktischen Uhrmacherkunst Seit der Altesten Art den Tag Einzutheilen Bis an das Ende des Achzehnten Jahrhunderts Johann Heinrich Moritz Poppe, Roch und Compagnie. Leipzig, Germany, 1801.
Avis Sur le Privilége des Horloges et des Montres de la Nouvelle Invention J. de Hautefeuille, Paris, France.
Clock and Watchmakers' Manual M. L. Booth. John Wiley, New York City, 1860.
Clock and Watchmakers' Manual, New and Complete Mary L. Booth. J. Wiley, New York City, 1860.
Clock and Watchmaking, Rudimentary Treatise on E. B. Denison (Lord Grimthorpe). John Weale, London, England, 1850.

Clock and Watchmaking, Treatise on Thomas Reid. Blackie and Son, London, England, 1849.

Clock and Watch Repairing, Essentials of John Drexler, Milwaukee, Wisconsin, 1914.

Clock and Watch Work From the Eighth Edition of the Encyclopedia Britannica Sir Edmund Beckett. Adam and Charles Black, 1855.

Clockjobber's Handybook, The Paul N. Hasluck. Crosby Lockwood and Son, London, England, 1899.

Clock, Watches and Bells Sir Edmund Beckett. (Sixth edition Revised and Enlarged.) Lockwood and Company, London, England, 1874.

Clockwork, Essays on the Improvement of Alexander Cumming, London, England, 1766.

Collection Archeologique du Prince Pierre Soltykoff. Horlogerie. Description et Iconographie des Instruments Horaires du XVIe Siècle, Précédée d'un Abrégé Historique de L'Horlogerie au Moyen Age Pierre Dubois. V. Didron, Paris, France, 1858.

Curiosities of Clocks and Watches E. J. Wood. R. Bentley, London, England, 1866.

Detached Lever Escapement, The Moritz Grossman. (Revised, Corrected, Enlarged.) Jewelers' Publishing Co., Chicago, Illinois, 1884.

Detached Lever Escapement A Discourse on The (Pamphlet.) C. T. Higginbotham. South Bend Watch Co., 1912.

Die Penduluhr Horologium Oscillatorium Christian Huyghens, 1673. W. Engelman, Leipzig, Germany, 1913.

English Trades, Book of Sir Richard Phillips. (Twelfth edition.) London, England, 1824.

Essai Sur L'Horlogerie, Relativement à L'Usage Civil, à l'Astronomie et à la Navigation 2 Vols., Paris, France, 1763.

Evolution of Automatic Machinery E. A. March. Geo. K. Hazlitt and Co., Chicago, Illinois, 1896.

Evolution of the Time-Piece Lyon and Scott. Ottumwa, Ohio, 1895.

Friction, Lubrication and Lubricants W. T. Lewis. Geo. K. Hazlitt and Co., Chicago, Illinois, 1896.

Geschichte der Uhrmacherkunst Emanuel Schreiber. B. Fr. Voigt, Weimar, Germany, 1850.

Great Industries of United States Horace Greeley. J. B. Burn, Hyde and Co., Chicago, Illinois, 1871.

Histoire Corporative de L'Horlogerie de L'Orfèvrerie et des Industries Annexes Anthony Babel. A. Kundig, Geneva, Switzerland, 1916.

Histoire de la Mésure du Temps par les Horloges Ferdinand Berthoud, Paris, France, 1802.

Histoire de L'Horlogerie Pierre Dubois. Published under management of "Moyen Age et la Renaissance," Paris, France, 1849.

History of Inventions, Discoveries and Origins Johann Beckman. Tr. from German by Wm. Johnston. Revised and Enlarged by Wm. Francis and J. W. Griffith, London, England, H. G. Bohn, 1846.

History of Watches and Other Timekeepers, A J. F. Kendal. Crosby Lockwood and Son, London, England, 1892.

Industrial History of the United States Albert Sidney Bolles. Henry Bill Publishing Co., Norwich, Connecticut, 1879.

Jewelled Bearings for Watches C. T. Higginbotham (Pamphlet.) G. K. Hazlitt and Co., Chicago, Illinois, 1911.

Journal Suisse D'Horlogerie Publié sous les auspices de la classe d'industrie et de commerce. (Société des arts de Genève.) 1876.

L'Art de Conduire et de Régler les Pendules Ferdinand Berthoud. Paris, France. 1805. 1811.

Les Montres Sans Clef Adrien Philippe. Geneva, Switzerland. 1863.

Lessons in Horology Jules Grossman and Herman Grossman. Keystone, Philadelphia, Pennsylvania, 1905.

Les Transformations Industrielles Dans L'Horlogerie Suisse Henri Borle. G. Krebs. 1910.

Lever Escapement, The T. J. Wilkinson. Technical Publishing Co., Philadelphia, Pennsylvania, 1916.

L'Horlogerie Astronomique et Civile; Ses Usages Ses Progrès Son Enseignement à Paris A. H. Rodanet. Vve. C. Dunod, Paris, France, 1887.

L'Industrie Horlogère aux États Unis George Blondel. Soc. de géographie commerciale de Paris. Bull. mensuel. Paris, France, 1917.

Manipulation of Steel in Watchwork John J. Bowman. Jewelers Circular Publishing Co., New York City, 1903.

Modern Clock, The Ward L. Goodrich. Hazlitt and Walker, 1905.

Modern Horology Claudius Lanier. Trans. by J. Tripplin. E. Rigg. (Second Edition.) Crosby Lockwood and Co., London, England, 1887.

Modern Horology, Treatise on Claudius Lanier. Translation.

Modern Methods in Horology Grant Hood. Kansas City, Jeweler and Optician, Kansas City, Missouri, 1904.

Nouveau Régulateur des Horloges des Montres et des Pendules; Ouvrage Mis à La Portée de Tout Le Monde et Orné de Figures Ferdinand Berthoud and L. Janvier, Paris, France, 1838.

Old Clock Book Mrs. N. Hudson Moore. Frederick A. Stokes and Co., New York City, 1911.

Old Clocks and Watches and Their Makers F. J. Britten. B. T. Batsford, London, England, 1899. Revised and Enlarged, 1914.

Old English Clocks F. J. Britten. Lawrence and Jellicoe. London, England, 1907.

Old Scottish Clockmakers John Smith. W. J. Hay, Edinburgh, Scotland, 1903.

Short Talks to Watchmakers C. T. Higginbotham. (South Bend Watch Co.) 1912. (Pamphlet.)

Simple and Mechanically Perfect Watch, A Moritz Grossman, G. K. Hazlitt and Co., Chicago, Illinois, 1891.

Sun-Dials, Book of Mrs. Alfred Gatty. Bell and Daldy, London, England, 1872.

Sun-Dials and Roses Alice Morse Earle. Macmillan Co., London, New York City, 1902.

Sur Les Anciens Horloges et Sur Jacques de Dondis Surnommé Horologius Falconet Camille. In Liber C. Col. D. V. 16, 1838.

Time and Clocks A Description of Ancient and Modern Methods of Measuring Time (Sir) H. H. Cunnynghame, M.A., C.B., M.I.E.E. Archibald Constable and Co., London, England, 1906.

Time and Its Measurement James Arthur. (Reprinted from Popular Mechanics Magazine.) Chicago, Illinois, 1909.

Time and Timekeepers L. and A. Mathey. (Pamphlet.) 1877.

Time and Timekeepers Adam Thomson. T. and W. Boone, London, England, 1842.

Time and Time Tellers J. W. Benson. Robert Hardwicke, London, England, 1875.

Timekeeper Invented by the Late Thomas Mudge, The By Thomas Mudge his son. Printed for the author, London, England, 1799.

Tower Clock and How to Make It E. B. Ferson. Hazlitt and Walker, Chicago, Illinois, 1903.

Universal Clock Adjuster Eleazar Thomas Perdue. Richmond, Virginia, 1877.

Watch, The Henry F. Piaget. Third edition. A. N. Whitehorne, New York City, 1877.

Watch Adjusters' Manual Charles Edgar Fritts. Charles E. Fritts, London, England, New York City, Toronto, Canada, 1894. (Third edition revised.)

Watch and the Clock, The Rev. Alfred Taylor. Phillips and Hunt, New York City, 1883.

Watch and Clock Escapements Keystone. Philadelphia, Pennsylvania, 1904.

Watch and Clockmaker's Handbook, Dictionary and Guide. F. J. Britten. E. and F. N. Spon, London. Spon and Chamberlain, New York City. (Tenth edition), 1902.

Watch and Clockmaking David Glasgow. Cassel and Co., Ltd., London, England; Paris, France; Melbourne, Australia, 1897.

Watch Balance and Its Jeweling, The (A lecture) C. T. Higginbotham. (South Bend Watch Co.) 1907.

Watch Factories of America, The Henry G. Abbott. Geo K. Hazlitt and Co., Chicago, Illinois, 1888.

Watchmaker and Machinists' Handbook Wm. B. Learned. G. K. Hazlitt and

Co., Chicago, Illinois, 1897.

Watchmakers' and Jewelers' Handbook C. Hopkins. John P. Morton and Co., Louisville, Kentucky, 1866.

Watchmakers' and Jewelers' Practical Handbook Henry G. Abbott. Fifth edition revised and enlarged. Geo. K. Hazlitt and Co., Chicago, Illinois, 1892.

Watchmakers' Handbook Claudius Lanier.

Watchmakers' Lathe W. L. Goodrich. Hazlitt and Walker, Chicago, Illinois, 1903.

Watchmakers' Tables The American Jeweler, Chicago, Illinois, 1914.

Watchmaking in America Reprint from Appleton's Journal. Robbins, Appleton and Co., 1870.

Watch Repairing F. J. Garrard. Crosby Lockwood and Son, London, England, 1903.

Watch Tests A Booklet of Tables F. M. Bookwalter, Springfield, Ohio, 1911.

Watchwork, Treatise on H. L. Melthropp, M.A., F.S.A. E. and F. M. Spon, London, England, 1873.

Worshipful Company of Clockmakers of London, The Catalogue of the Museum of Second edition. Blades, East and Blades, London, England, 1902.

Workshop Notes for Jewelers and Watchmakers Compiled by Charles Brassler. Jewelers' Circular Publishing Co., New York City, 1892.

APPENDIX C

American Watch Manufacturers
(CHRONOLOGY)

Judged by the number of failures which have marked the development of the American watch industry, watch manufacturing might well be characterized as a perilous business. While it has proved profitable for a few, it also has swallowed many fortunes.

There were no watch companies in America until 1850, although a few attempts were made to manufacture watches in the United States prior to that time by Luther Goddard, who established the first American watch factory at Shrewsbury, Massachusetts, in 1809 and made several hundred watches from 1809 to 1815, when he finally abandoned the business; by Henry and James F. Pitkin at East Hartford, Connecticut, from 1838 to about 1845 and by Jacob D. Custer at Norristown, Pennsylvania, from 1840 to 1845.

Except for a few companies whose organization and speedy dissolution had small, if any, effect upon the industry as a whole, the following briefly outlines the history of American watch manufacturing companies from the real beginning in 1850 to the present day:

1850

The American Horologe Company of Roxbury, Massachusetts, organized; name changed same year to The Warren Manufacturing Company; in 1853 name was again changed to The Boston Watch Company, the principal stockholders of which organized The Waltham Improvement Company to buy land and buildings for The Boston Watch Company at Waltham, Massachusetts; moved into the new factory at Waltham in 1854; failed in 1857 and company's business was bought in by Royal E. Robbins, watch importer of New York City and Tracy and Baker, watch case manufacturers of Philadelphia; in 1858 The Waltham Improvement Company increased its capital and purchased the business and property of The Boston Watch Company and re-incorporated under the name of The American Watch Company; in 1885 the name was changed to The American Waltham Watch Company and in 1906 the name was again changed to The Waltham Watch Company, its present name; in 1913 the Company purchased the business of the Waltham Clock Company.

1857

E. Howard and Company of Roxbury, Massachusetts, was organized by Edward Howard; in 1861 the name was changed to The Howard Clock and Watch Company; in 1863 the company practically failed and was reorganized under the name of The E. Howard Watch and Clock Company; in 1881 the Company again practically failed and was again reorganized under the name of The E. Howard Watch and Clock Company, with Edward Howard as President, as he had been in the preceding organizations; in 1882 Howard withdrew as President and severed his connection with the Company. From that time forward the Company gave increasingly greater attention to the manufacture of clocks, although it continued to manufacture the Howard watch until about 1903 when it entered into a contract with The Keystone Watch Case Company of Philadelphia, under which The E. Howard Watch and Clock Company transferred to The Keystone Company all rights to the use of the name "E. Howard" in connection with the manufacture of watches and also changed its own corporate name to The E. Howard Clock Company. Later the company failed and was operated by receivers until 1910 when a new company of the same name was organized and purchased the property of the old concern. The Keystone Company purchased the factory of The United States Watch Company at Waltham, Massachusetts, and began the manufacture of watches under the name of The Howard Watch Company.

1859

The Nashua Watch Company of Nashua, New Hampshire, was organized; it failed in 1862 and was bought in by the American Watch Company now The Waltham Watch Company.

1863

The Newark Watch Company of Newark, New Jersey, was organized; it sold out to The Cornell Watch Company of Chicago in 1870.

The United States Watch Company of Marion, New Jersey, was organized; it failed in 1872 and was operated by creditors for a short time under the name of The Marion Watch Company, but again failed; machinery of the company was sold to E. F. Bowman of Lancaster, Pennsylvania, who manufactured a few watches and then sold the business to The J. P. Stevens Watch Company of Atlanta, Georgia.

1864

The National Watch Company was organized and erected a factory at Elgin, Illinois; in 1874 the name was changed to its present name of The Elgin National Watch Company.

The Tremont Watch Company of Boston was organized, with Aaron L. Dennison, one of the founders of the original Waltham Watch Company as superintendent; it ceased business in 1868 because of lack of capital; machinery of the company was sold to an English syndicate which organized in England The Anglo-American Watch Company, the name of which was later changed to The English Watch Company.

The New York Watch Company of Springfield, Massachusetts, was organized by Don J. Mozart and others; it practically failed in 1866 and was reorganized under the same name; again failed in 1870 and the business was taken over by a new company known as The New York Watch Manufacturing Company. This Company survived only a few months and the property and business were taken over by a new group in January 1877 under the name of The Hampden Watch Company, which company, in turn, was later purchased by John C. Deuber and associates in control of The Deuber Watch Case Manufacturing Company of Canton, Ohio, which was originally organized at Cincinnati about 1888.

1867

The Mozart Watch Company of Ann Arbor, Michigan, was organized by Don J. Mozart after leaving The New York Watch Company; in 1871 the property and business were sold to The Rock Island Watch Company of Rock Island, Illinois.

1869

The Illinois Springfield Watch Company was organized; in 1875 it was reorganized under the same name; in 1879 it was again reorganized and the name was changed to The Springfield Illinois Watch Company, which was later changed to The Illinois Watch Company, under which name it now operates.

1870

The Cornell Watch Company of Chicago was organized and took over the business of The Newark Watch Company of Newark, New Jersey; in 1874 it sold its business and property to The Cornell Watch Company of San Francisco, California.

1871

The Rock Island Watch Company of Rock Island, Illinois, was organized and purchased the business of The Mozart Watch Company of Ann Arbor, Michigan; it failed the same year without producing any watches and passed out of existence.

1872

The Washington Watch Company of Washington, D. C., was organized, but failed after two years.

1873

The Rockford Watch Company of Rockford, Illinois, was organized; in 1896 the company failed and the business was operated by assignee until 1901 when it was sold and reorganized under the name of The Rockford Watch Company, Ltd.; it discontinued business in 1915, since which time the remaining stock has been marketed by The Illinois Watch Case Company of Elgin, Illinois.

1874

The Adams and Perry Watch Manufacturing Company of Lancaster, Pennsylvania, was organized; it failed in 1876 without producing any watches; the property was purchased by a syndicate in 1877 which organized under the name of The Lancaster Pennsylvania Watch Company; in 1878 it was reorganized under the name of The Lancaster Pennsylvania Watch Company, Limited; in 1878 it was again reorganized under the name of The Lancaster Watch Company. In 1884 control of the company passed to Abram Bitzner, who, with Oppenheimer Bros. and Vieth, selling agents of New York City, began to operate the company and assumed the name of "Keystone Watch Company" as a trade mark; they failed in 1890 and in 1892 the property was purchased by The Hamilton Watch Company.

The Freeport Watch Manufacturing Company of Freeport, Illinois, was organized, but before producing any watches the company's factory burned and the business was discontinued in 1875.

1874

The Cornell Watch Company of San Francisco, California, was organized and took over the business of the Cornell Watch Company of Chicago; in 1875 the company was reorganized under the name of The California Watch Company and in 1877 the business was sold to the Independent Watch Company of Fredonia, New York.

1875

Fitchburg Watch Company of Fitchburg, Massachusetts, was organized, but discontinued, for lack of funds, a few years later without producing any watches.

1877

The Hampden Watch Company, now of Canton, Ohio, was organized at Springfield, Massachusetts and took over the business of the New York Watch Company; later, the Company's business and property were purchased by the interests in control of the Deuber Watch Case Manufacturing Company of Canton, Ohio.

The Independent Watch Company of Fredonia, New York, was organized and purchased the business and property of the California Watch Company of San Francisco; in 1885 the business was sold to the Peoria Watch Company of Peoria, Illinois.

1879

The Auburndale Watch Company, of Auburndale, Massachusetts, was organized and purchased the machinery of the United States Watch Company of Marion, New Jersey. In 1883 the company made a voluntary assignment.

1880

The Waterbury Watch Company of Waterbury, Connecticut, was incorporated; in 1898 the name of the company was changed to the New England Watch Company; in 1912 the company failed, and in 1914 the property was sold to and is now operated as one of the factories of Robt. H. Ingersoll and Brothers. of New York City.

The E. Ingraham Company of Bristol, Connecticut, founded by E. Ingraham in 1835 for the manufacture of clocks, was incorporated; in 1912 the company purchased the business of The Bannatyne Watch Company of Waterbury, Connecticut.

The Western Watch Company of Chicago was organized but failed the same year without producing any watches, the machinery being sold to The Illinois Watch Company.

1882

The Columbus Watch Company was organized at Columbus, Ohio; it was the outgrowth of a private enterprise started in 1876 by D. Gruen and W. J. Savage, who imported watch movements from Switzerland and sold them in American-made cases. In 1903 the business of the company was purchased by The South Bend Watch Company of South Bend, Indiana.

The J. P. Stevens Watch Company of Atlanta, Georgia, was organized and failed in 1887.

1883

The New Haven Watch Company of New Haven, Connecticut, was organized; in 1886 the company moved to Chambersburg, New Jersey, then a suburb of Trenton; in the same year the name of the company was changed to The Trenton Watch Company; in 1907 the company failed and in 1908 the business and property were acquired by Robt. H. Ingersoll and Brothers. of New York City. The factory at Trenton has since been operated as one of the plants of the Ingersolls.

The Manhattan Watch Company of New York City was organized but did not long continue.

The Cheshire Watch Company of Cheshire, Connecticut, was organized and continued in operation for about ten years.

The Aurora Watch Company of Aurora, Illinois, was incorporated but did not begin operations until 1885; failed in 1886; machinery sold in 1892 to The Hamilton Watch Company of Lancaster, Pennsylvania.

1884

The Seth Thomas Clock Company of Thomastown, Connecticut, founded by Seth Thomas in 1813 and incorporated in 1853, began the manufacturing of watches in 1884, but discontinued their manufacture in 1914. Seth E. Thomas, Jr., great-grandson of the founder, is now president of the company.

The United States Watch Company of Waltham, Massachusetts, was organized as an outgrowth of The Waltham Watch Tool Company. Later it failed and its plant was purchased by The Keystone Watch Case Company, which operates the factory under the name of The Howard Watch Company.

1885

The New York Standard Watch Company of Jersey City, New Jersey, was organized; in 1902 it was purchased by The Keystone Watch Case Company, which continues to operate it under the original name.

The Peoria Watch Company of Peoria, Illinois, was organized and took over the business of The Independent Watch Company of Fredonia, New York, but did not long survive.

1887

The Wichita Watch Company of Wichita, Kansas, was organized, but continued in operation only a few years.

1888

The Western Clock Manufacturing Company was incorporated with factory at Peru, Illinois, and general offices at La Salle, Illinois; began manufacturing watches in 1895; in 1895 the name of the company was changed to Western Clock Company; manufacturers of "Big Ben" alarm clock and low-priced nickel watches.

1890

D. Gruen Sons and Co., of Cincinnati, originally incorporated under laws of West Virginia; in 1898 re-incorporated under laws of Ohio. Prior to original incorporation the business was operated as a partnership under the name of D. Gruen and Sons. Present company also operates under the trade name of Gruen Watch Case Co. The company manufactures its watch movements in Switzerland, assembling and casing them in the United States.

1892

The Hamilton Watch Company of Lancaster, Pennsylvania, was organized; made only movements until 1909, but since then, both cases and movements.

1893

Robt. H. Ingersoll and Bro., of New York City, first introduced the original Ingersoll watch to the public at the World's Columbian Exposition; in 1892 the Ingersolls had contracted with the Waterbury Clock Company of Waterbury, Connecticut for the manufacture of the low-priced watch, which was first sold for $1.50 and later for $1.00; in 1908 the Ingersolls purchased the factory and business of the Trenton Watch Company of

Trenton, New Jersey, and began watch manufacturing on their own account; in 1914 they purchased the plant of The New England Watch Company, formerly The Waterbury Watch Company of Waterbury, Connecticut.

1894

The Webb C. Ball Company of Cleveland, Ohio, founded in 1879 and incorporated in 1891, began the manufacture of watches.

1899

The Keystone Watch Case Company of Philadelphia, Pennsylvania, was organized. It controls The Howard Watch Company of Waltham, Massachusetts, The New York Standard Watch Company of Jersey City, New Jersey, The Crescent Watch Case Company, Inc., of Newark, New Jersey, and The Philadelphia Watch Case Company of Riverside, New Jersey.

1902

The South Bend Watch Company of South Bend, Indiana, was incorporated in New Jersey under the name of The American National Watch Company, but immediately thereafter changed to its present name; in 1903 it purchased the business of The Columbus Watch Company of Columbus, Ohio; in 1913 it was re-incorporated under Indiana laws.

1904

The Ansonia Clock Company of Brooklyn, New York, incorporated in 1873, began the manufacture of low-priced nickel watches; its principal business, however, is that of clock manufacture.

1911

The Leonard Watch Company of Boston, Massachusetts, was incorporated for the purpose of selling and distributing watches.

APPENDIX D

Well-Known Watch Collections
(From list compiled by Major Paul M. Chamberlain, of Chicago in 1915.)
Abbott George E. H. Abbott, Groton, Massachusetts.
Addington S. Addington, Esq., purchaser at Bernal sale.
Ashmolean Ashmolean Museum, Oxford, England.
Augsburg Maxmillian Museum, Augsburg, Germany.
Baker Edwin P. Baker, referred to by Britten.
Baxter James Phinney Baxter, Portland, Maine.
Blois Musee de la ville, Blois, France.
Boston Museum of Fine Arts, Boston, Massachusetts.
Bourne T. W. Bourne, referred to by Britten.
British British Museum, London, England.
Bulley Edward H. Bulley, referred to by Britten.
Burkhardt M. Albert Burkhardt, Basle, Switzerland.
Chamberlain Paul M. Chamberlain, Chicago, Illinois.
Chesam Lord Chesam, referred to by Britten.
Cluny Musee de Cluny, Paris, France.
Clarke A. E. Clarke, London, England.
Cockey Edward C. Cockey, New York City.
Cointre La Famille Cointre, of Poitiers, France.
Copenhagen Horological Museum, Copenhagen, Denmark.
Cook E. E. Cook, Walton-on-Thames, England.
Czar Imperial collection, Hermitage Gallery, Petrograd, Russia (1915).
Cumberland Duke of Cumberland, England.
Debruge Debruge collection, catalogue published in 1849, referred to by M. E. Deville in Les Horlogers Blesois.
Dennison Franklin Dennison collection, Birmingham, England.

Devotion The Edward Devotion House, Brookline, Massachusetts.

Dickson R. Eden Dickson, London, England.

Ditisheim Henri Ditisheim, Chaux-de-Fonds, Switzerland.

Dresden Green Vaulted Chambers, Dresden, Germany.

Duplessis Family of Duplessis of Blois, referred to in Les Horlogers Blesois.

Dover Dover Museum, Dover, England.

Dunwoody Dr. W. J. Dunwoody, mentioned by Britten.

Estreicher Dr. Tad. Estreicher, Fribourg, Switzerland.

Eschenbach Baroness Marie von Ebner-Eschenbach, Vienna, Austria-Hungary.

Fawkes J. H. Fawkes of Farnlet Hall, England.

Fellows Collection of Sir Charles Fellows, of Westbourn, Isle of Wight, bequeathed by widow to British Museum.

Fitzwilliam Fitzwilliam Museum, Cambridge, England.

Fleisher Collection of Moyer Fleisher, exhibited in the Pennsylvania Museum,

Memorial Hall, Philadelphia, Pennsylvania.

Foulc M. Foulc, Paris, France.

Franck B. Bernard Franck, Paris, France.

Freeman Charles Freeman, referred to by Britten.

Froidevaux M. Froidevaux, Blois, France.

Garnier M. Paul Garnier, Paris, France.

Gelis M. Edouard Gelis, Paris, France.

Geyer H. F. Geyer, mentioned by Britten.

Georgi M. Georgi, Paris, France.

Glyn George Carr Glyn, referred to by Britten.

Gotha Museum of Gotha, Germany.

Greene T. Whitcomb Greene, referred to by Britten.

Guildhall Guildhall Museum, London, England.

Hartshorne Albert Hartshorne, referred to by Britten.

Hearn George Hearn collection, presented by widow to Metropolitan Museum of Art, New York City.

Heckscher Martin Heckscher collection in Vienna, Austria-Hungary.

Heinz Collection of Henry J. Heinz, exhibited in the Carnegie Museum, Pittsburg.

Hodgkins Collection of J. E. Hodgkins, London, England.

Humphreys Miss M. Humphreys, mentioned in Britten.

Jenkins Collection of Jefferson D. Jenkins, Decatur, Illinois.

King C. King, Newport, Monmouthshire, England.

Kensington South Kensington Museum, London, England.

Kirner B. A. Kirner, Chicago, Illinois.

Lambert Messrs. Lambert, referred to by Britten.

Lazerus Collection of Moses Lazerus, Philadelphia, bequeathed to Pennsylvania Museum, Philadelphia, Pennsylvania.

Lambiley Compte de Lambiley, France.

Laurance E. A. Laurance, mentioned by Britten.

Lebenheim Mentioned in Morgan catalogue.

Lecointre Family of Lecointre, Poitiers, France.

Leicester Leicester Museum, Leicester, England.

Leroux M. E. Leroux, Paris, France.

Liljigren L. O. Liljigren, Chicago, Illinois.

Londesboro Lord Londesboro, London, England.

Louvre Musee de Louvre, Paris, France.

Marfels Collection of Carl Marfels, Berlin, Germany.

Massey Edwards Massey, London, England.

Meldrum Robert Meldrum, referred to by Britten.

Metropolitan Metropolitan Museum of Art, New York City.

Mirabaud M. G. Mirabaud, Paris, France.

Moore Bloomfield Moore collection in Pennsylvania Museum, Philadelphia.

Morgan J. Pierpont Morgan collection at Metropolitan Museum of Art, New York City.

O. Morgan Octavius Morgan collection in British Museum.

Moray Lord Moray, London, England.

Moss Rev. J. J. Moss, purchaser at Bernal sale, London, England, 1855.

Munich National Bavarian Museum at Munich, Germany.

Nelthropp Collection presented by Rev. H. L. Nelthropp to the Worshipful Company of Clockmakers of the City of London and exhibited at Guild Hall Museum.

Newington Newington Free Library, Newington, England.

Olivier M. Olivier, Paris, France.

Parr Edward Parr, London, England.

Partridge R. W. Partridge, London, England.

Ponsonby Hon. Gerald Ponsonby, referred to by Britten.

Proctor Frederick Towne Proctor, Utica, New York.

Proctor, T. R. Thomas Redfield Proctor, Utica, New York.

Purnell J. B. Purnell, purchaser at Bernal sale in 1855.

Ranken William Ranken, London, England.

Reeves R. F. Reeves, St. Louis, Missouri.

Renouard Family of Renouard, Belois, France.

Roberts Evan Roberts, London, England.

Robertson J. Drummond Robertson, London, England.

Roblot Ch. Roblot, Paris Passy, France.

Rothchild Baroness Alphonse de Rothchild collection.

Rosenheim Max Rosenheim, referred to by Britten.

Roux Edward Roux, mentioned by Britten.

Salting Collection now in the South Kensington Museum.
Saussure M. Th. de Saussure, mentioned by Britten.
Sauve M. Sauve, Belois, France.
Schlichting Baron von Schlichting, Petrograd, Russia, (1915).
Shapland Charles Shapland, London, England.
Shaw Morgan Shaw, London, England.
Sidebottom Collection of Mrs. H. Sidebottom, in South Kensington Museum.
Sivan M. Charles Sivan, Paris, France.
Smythies Major R. H. Raymond Smythies, London, England.
Soane Soane Museum, London, England.
Stamford Stamford Institution, England.
Stroehlin Stroehlin collection, referred to in J. P. Morgan catalogue.
Sudell Edward Sudell, mentioned by Britten.
Sutton Rev. A. F. Sutton, England.
Thompson Mrs. G. F. Thompson, Ottawa, Canada.
Torphicon Lord Torphicon, referred to by Britten.
Turrettini Turrettini collection referred to by Dr. Williamson in Morgan catalogue.
Vautier M. L. Vautier, Belois, France.
Vendome Calvaire de Vendome, France.
Vienna Imperial Treasury, Vienna, Austria-Hungary.
Wallace Lord Wallace collection, bequeathed by his widow to the British Museum.
Wehrle Eugene Wehrle, Brussels, Belgium.
Wheeler, H. L. Horace L. Wheeler, Boston, Massachusetts.
Wheeler Collection of Willard H. Wheeler, Brooklyn, N. Y., exhibited in the Brooklyn Museum, New York City.

APPENDIX E

Encyclopedic Dictionary

Abrasion Wearing away by rubbing or friction.

Adams, J. C. A promoter instrumental in organizing the Elgin, Illinois, Cornell, and Peoria Watch Companies, and the Adams and Perry Manufacturing Company. He invented and patented the "Adams System" of time records in use on most of the railroads in the West. He last appeared in prominent connection with the watch and clock business as the organizer of the Swiss horological exhibit at the World's Columbian Exposition.

Addenda Tips of the teeth of a wheel beyond the pitch circle. Sometimes of circular outline; sometimes ogive that is, of a shape patterned after the pointed arch. The addendum is also known as the "face" of the tooth.

Adjustment The manipulation of the balance with its spring and staff to secure the most accurate time-keeping possible. Three adjustments are usually made, viz.: for isochronism, temperature and position. Much of the difference in value and cost of watches depends on this operation.

Adjustment to Isochronism Strictly speaking this would cover all adjustment; but it is technically understood to mean an adjustment of the balance spring so that the time of vibration through the long and short arcs of the balance is the same.

Adjustment to Positions The manipulation of the balance and its spring so that a watch keeps time in different positions. Good watches are usually adjusted to five positions. They are pendant up; III up; IX up; dial up; and dial down.

Adjustment to Temperature or Compensation The adjustment of the balance and spring so that the time-keeping qualities are affected as little as possible by changes in temperature. See Compensation.

Ahaz King of Judea, 742-727 B. C. See Dial of Ahaz.

Alarm Sometimes spelled "alarum." A mechanism attached to a clock whereby at any desired time a bell is struck rapidly by a hammer.

Aluminum-Bronze An alloy of aluminum and pure copper, usually in the proportion of 10 parts of the former and 90 of the latter. It is considerably lighter than brass and highly resistant to wear.

Anaximander Greek astronomer to whom the Greeks ascribed the invention of the sun-dial in the sixth century B. C.

Arbor The axle or axis on which a wheel of a watch or clock turns. Also applied to a spindle used by watchmakers.

Arc Any section of the circumference of a circle.

Archimedes A famous Greek philosopher and scientist sometimes credited with the invention of the clock. About 200 B. C. he made a machine with wheel work and a maintaining power but having no regulator it was no better as a time teller than a planetarium turned by a handle. It may have furnished the suggestion for later time-keeping machines.

Arnold, John Born 1736. An English watchmaker of note. He invented the helical form of the balance spring and a form of chronometer escapement much like Earnshaw's. Died 1799. Arnold's devices have been most useful and permanent.

Assembling The putting together of the finished parts of a watch. In a three-quarter plate watch this is done on the lower plate. In a full plate movement it is easier and more satisfactory to assemble on the top plate.

Astrolabe 1. An instrument of various forms formerly used especially in navigation to measure the altitudes of planets and stars. 2. A projection of a sphere upon any of its great circles.

Astronomical Time Means solar time, as computed from observing the passage of the sun across the meridian from noon of one day to noon of the following day. It is counted continuously up to 24 not in two 12-hour divisions.

Astronomy The science which treats of the motions, real and apparent, of the heavenly bodies. Upon this science, through its determination of the length of the year, is founded the science of horology or time-keeping.

Automata for Striking Very common on old clocks and very complicated, such as: Indian King hunting with elephants, Adam and Eve, Christ's flagellation, and many others. See Clocks, Interesting Old.

Automatic Machinery The second great contribution of America to watchmaking after the establishment of the principle of interchangeability of parts, and making possible the effective execution of that principle.

Auxiliary A device attached to a compensation balance to reduce what is known as the "middle temperature error." Some are constructed to act in high temperatures only as Molyneux's; and some in low temperatures only as Poole's.

Balance The vibrating wheel in a watch or chronometer which with the aid of the balance spring (hair-spring) regulates the rate of travel of the hands. The balance is kept in vibration by means of the escape wheel. See Compensation Balance.

Balance Arc In detached escapements, that part of the vibration of the balance in which it is connected with the train. The remainder is called the drop.

Balance-Clock A form of clock built before the pendulum came into use. The regulating medium was a balance on the top of the clock made with a verge escapement. See Foliot.

BALANCE COCK

Balance Cock The standard which supports the top pivot of the balance. In old watches often elaborately pierced and engraved.

Balance Spring In America usually called the "hair-spring." A long slender spring that governs the time of vibration of the balance. One end of the balance spring is fastened to a collet fitted friction-tight on the balance staff, the other to a stud attached to the balance cock or to the watch plate. The most ordinary form is the volute, or flat spiral. The other form used is an overcoil. See Bréquet Spring. The principle of the isochronism of a balance spring was discovered by Hooke, and first applied to a watch by Tompion. The name hair-spring comes from the fact that the first ones are said to have been made from hog bristles.

Balance Spring Buckle or "Guard" A small stud with a projecting tongue attached to the index arm and bridging the curb pins so as to prevent their engaging two of the balance spring coils. Used chiefly in Swiss watches.

Balance Staff The axis of the balance. The part of a watch most likely to be injured by a fall.

Balance Wheel A term often incorrectly applied to the balance itself, but properly it is the escape wheel of the verge escapement.

Band Of a Watchcase The "middle" of the case to which the dome, bottom and bezel are fastened; the last sometimes screwed, sometimes snapped.

Bank Banking-pin.

Banking In a lever watch the striking of the outside of the lever by the impulse pin due to excessive vibration of the balance. In a cylinder or verge movement the striking of the pin in the balance against the fixed banking-pin.

Banking-Pin A pin for restricting the motion of the balance in verge and cylinder watches.

Banking-Pins 1. In a lever watch, two pins which limit the motion of the lever. 2. In a pocket chronometer, two upright pins in the balance arm which limit the motion of the balance spring. 3. In any watch, the curb pins which confine the balance spring are sometimes called banking-pins.

Barlow, Edward (Booth) A clergyman of the Church of England, born in 1636. He devoted a great deal of time to horological pursuits. He invented the rack repeating striking works for clocks, applied by Tompion in 1676. He invented also a repeating works for watches on the same plan. And he invented the cylinder escapement which he patented with Tompion and Houghton. When he applied for a patent on his repeating watch he was successfully contested by Quare, who was backed by the Clockmakers' Company. He died in 1716.

Bar Movement A watch movement in which bars take the place of the top plate and carry the upper pivots. Sometimes termed a "skeleton" movement. Not generally adopted because its many separate bearing parts promote inaccuracies where large quantities are to be produced.

Barrel A circular box which confines the mainspring of a watch or clock.

Barrel Arbor The axis of the barrel around which the mainspring is coiled.

Barrel Hollow A sink cut either into the top plate or the pillar plate of a watch to allow the barrel freedom.

Barrel Hook A bent pin in the barrel to which the mainspring is attached.

Barrel Ratchet A wheel on the barrel arbor which is prevented by a dog from turning backward while the mainspring is being wound and which becomes the base against whose resistance the train is driven.

Bartlett, P. S. One of the early watchmakers of America. Connected with the Waltham factory at first and later with the Elgin Company. It is said that he first proposed the formation of the company at Elgin. His name became familiar as a household word throughout the country from being inscribed upon a full-plate model which attained widespread success.

Beat The strike or blow of the escape wheel upon the pallet or locking device.

Beat Pins The pins at the ends of the pallets in a gravity escapement which give impulse to the pendulum.

Beckett, Sir Edmund See Denison, Edmund Beckett.

Berosus A Chaldean historian who lived at the time of Alexander the Great, about 200 B. C., and was a priest of Belus at Babylon. Said to have been the inventor of the hollow sun-dial. He was the great astronomer of his age.

Berthoud, Ferdinand, 1727-1807 An eminent French watchmaker and writer on horological subjects. Among his books are: "Essai sur l'Horlogerie," "Traite des Horloges Marines," and "Histoire de la mesure du Temps." He was a Swiss by birth, but lived most of his life in Paris.

Bezel The ring of a watch or clock case which carries the glass or crystal in an internal groove.

Big Ben The great bell which strikes the hours on the clock at Westminster.

Bizzle A corruption of Bezel. See Bezel.

Blow Holes Places where the brass and steel of a compensation balance are not perfectly united, when they are put together with silver or solder.

Bob The metal mass forming the body of a pendulum.

Boethius, Ancius Manlius Severinus, A. D. 480-524 A Roman philosopher and statesman to whom is sometimes attributed the invention of the clock. He did make a sun-dial and a water clock which latter may have contained a germ of the idea later developed into our modern clock.

Boss A cylindrical prominence or stud. The minute hand is carried on the boss of the center wheel.

Bottom Of a Watchcase The cover outside the dome of the case. Commonly called the "back."

Bouchon The hard brass tubing of which pivot holes in watch and clock plates are made; known commonly as "bushing wire." The short sections cut off for a pivot being called the "bushing."

Bow The ring of a watch case to which the guard or chain is attached; also known as "pendant bow."

BOW AND BUTTON

Box Chronometer A marine chronometer.

Boxing-In Fitting the watch movement in its case; applied chiefly to the encasing of stem-winding movements.

Bréquet, Abraham Louis A celebrated Swiss mechanician and watchmaker born at Neufchatel in 1747. He made several improvements in watches, the most notable being the Bréquet hairspring still in use in the best watches. He died in 1823.

BRÉQUET SPRING

Bréquet Spring A form of balance spring which is a volute with its outer end bent up above the plane of the body of the spring and carried in a long curve towards the center near which it is fixed. Like all other springs in which the outer coil returns towards the center, it offers opportunities of obtaining isochronism by varying the character of the curves described by the outer coil and thus altering its resistance. So-called from its inventor, Abraham Louis Bréquet (q. v.). Its advantage over the flat spring is that the overcoil allows expansion and contraction in all directions, thereby avoiding a good deal of side friction on the pivots as well as insuring more nearly perfect isochronism in changes of temperature.

Bridge A standard fastened to the plate, in which a pivot works.

Bridge Model The term given to watch movements in which plates or bridges carrying the upper pivots of the train rest firmly on the lower or dial plate and are held rigid by steady pins on lower side of the plate; the bridge being secured direct to the dial plate by screws termed plate or bridge screws. This is the most common construction of present-day manufacture and is utilized in three-quarter plate or separate and combination bridges covering one or more pivots of train wheels. Its alternate is "pillar model."

Buck, D. A. A. A watch repairer in Worcester, Mass., who designed a model for the Waterbury watch. His first model was not successful, but in 1877 he completed one which, a little later, the Waterbury Company, with Buck as master watchmaker, started to make. He remained with the company until 1884.

Bush A perforated piece of metal let into a plate to receive the wear of pivots.

Butting The engaging of the tips of the teeth of two wheels acting in gear. The proper point of contact being in the line of the shoulders of the teeth, butting is remedied by setting the wheels farther apart.

Button The milled knob used for winding and setting a keyless watch.

Calculagraph Trade name for a device for automatically computing and recording elapsed time in connection with factory jobs and other work where it is necessary to show the amount of labor used.

Calendar A system of dividing the year into months and days. The principal calendars known to history are: the Julian calendar; the Gregorian calendar; the Hebrew calendar; the Mohammedan calendar; and the Republican calendar. None of them has been quite accurate in dividing up the solar year, and frequent arbitrary corrections are necessary to secure a practical approximation. See descriptive article under each title.

Julian Established by Julius Caesar, 46 B. C., to remedy existing defects in the Roman calendar then in use. The Julian year was based on the assumption that the solar year is 365¼ days which was 11 minutes and 14 seconds too long. The scheme adopted was to make the regular calendar year 365 days, and to add one day every fourth year. The Julian calendar is still in use by Russia and Greece, where the dates now differ from those of most other countries by 13 days.

Gregorian Established October 15, 1582, by Pope Gregory XIII, in correction of the obvious errors of the Julian calendar. It is the calendar now in use by nearly all civilized nations. The mean length of the Gregorian year is 365 days, 5 hours, 49 minutes and 12 seconds 26 seconds longer than the actual solar year. Correction is made by adding a 29th day for February every fourth year, excepting when the date of said fourth year is divisible by 100. If, however, the date is also divisible by 400, the extra day is added.

Republican The calendar of the French Revolution (1793) declared to begin at midnight on the meridian of the Paris Observatory preceding the true autumnal equinox, September 22, 1792. There were 12 months of 30 days each and 5 or 6 "extra days" (as might be necessary) at the end of the year to bring the new year nearest to the then position of the equinox. Abolished

January 1, 1806.

Hebrew Composed of 12 lunar months, a thirteenth month being added from time to time to secure correspondence of the months with the passing seasons. The months are arbitrarily arranged to have alternately 29 days and 30 days. The length of the calendar year varies from 353 days to 385 days.

Mohammedan Based on a lunar year of 354 days divided into 12 lunar months which are alternately 29 and 30 days in length. During each period of 30 years a total of 11 days are added one at a time at the end of a year. The lack of co-ordination with the solar year results in a total separation of the seasonal year and the calendar year. In use in Turkey and some other Mohammedan countries.

CALENDAR CLOCK

Calendar Clock, or Watch A clock or watch which indicates days and months as well as hours.

Caliper The scheme of arrangement of a watch train, or the disposition of the parts of a watch.

CAM

Cam A rotating piece either non-circular or eccentric, used to convert rotary into linear reciprocating motion, oftener irregular in direction, rate, or time.

Cannon Pinion The pinion to which the minute hand is attached. It is tubular in form (whence its name), the main arbor passing through it friction-tight.

Canton Berne The Swiss district which does the largest export business in silver and base metal watches in Switzerland. The cantonal government has done everything possible to promote the industry, among other things: 1. Established information offices in the principal watch-making centers. 2. Established a permanent exhibition of articles used in the industry. 3. Established schools and associations and protective territories. 4. Prepared statistics and means for negotiating commercial relations.

Cap The part of the case that covers the movement.

Capped Jewel A jewel having a protective end-stone.

Carillon Chimes frequently used in the earlier clocks for striking the hours. Still used in some clocks.

Caron, Peter Augustus A famous Paris watchmaker, afterward called Beaumarchais, who made the first keyless watch of which we have any account.

Case The metal box in which the movement of a watch is inclosed.

Case-Springs The springs which cause the outer bottom of a watch case to fly open when the lock spring is released.

Center of Gyration That point in which the whole mass of a rotating body might be concentrated without altering its moment of inertia.

Center of Oscillation That point in a pendulum at which, if the whole mass of the pendulum were collected, the time of oscillation would be the same.

Center Seconds or Sweep Seconds A long seconds hand moved from the center of a watch dial, as are the minute and hour hands.

Center Staff The arbor attached to the center wheel which carries the minute hand.

Center Wheel The wheel in ordinary clocks and watches placed in the center of the frame on whose arbor the minute hand is carried. It is intermediate between the barrel and the third wheel.

Chamfer To cut away to a bevel the right angle formed by two adjacent faces as of a jewel or stone. It is also occasionally used to signify channeling or grooving.

CHAMFER

Chasing A form of ornament for metals which is made by punching or pressing from behind to present the pattern in relief instead of by cutting away the material.

Chops In a pendulum clock the blocks, usually of brass, between which the top of the pendulum suspension spring is clipped to prevent its twisting as it swings.

Chronograph In general, a recording clock or watch. Specifically, a watch with a center-seconds hand which may be stopped, started or returned to zero at will by pressing a button. Used for timing races, or measuring other short spaces of time with great exactness.

Chronometer Any very accurate time-keeper. Usually understood to mean a time-keeper fitted with a spring detent escapement. They usually have a fusee and a cylindrical balance spring.

Chronometer, Marine Probably the most exact form of time-keeper, especially for use on shipboard. The driving power is a mainspring acting by a chain on a fusee, and governed by what is known as the Chronometer or Detent Escapement, with, as a rule, the cylindrical balance spring. The movement is mounted on gimbals in an air and water-tight brass case, maintaining the dial constantly in a horizontal position.

Chronoscope A clock or watch in which the time is shown by figures presented at openings in the dial.

Church, Duane H. Credited with having contributed more to the automatic features of watch machinery than any other man. He was born in Madison County, N. Y., in 1849. At 16 he was apprenticed to a watchmaker of St. Paul, Minn., and after working at the trade for 17 years, he became in 1882 the master watchmaker for the Waltham Watch Company. Besides his invaluable contributions to automatic machinery, he improved the general design of watch movements and invented a form of pendant setting which enables stem-winding movements to be set in cases not especially adapted to them. He died in 1905.

Circular Error The difference in time arising from the swinging of a pendulum in a circular arc instead of its true theoretical path which is a cycloidal arc. This caused much trouble in the early clocks. Huyghens attempted to correct it (see Huyghens' Checks) but found that his device caused greater error. With the heavier pendulum and shorter arcs of vibration this error becomes negligible. The suspension of the pendulum by a flat flexible spring instead of a cord, attributed to Dr. Hooke, served to make the path practically cycloidal.

Cleopatra's Needle An Egyptian obelisk at whose base a dial was marked. Now in London. Another similar obelisk from Egypt is in Central Park, New York City.

Clepsammia The sand-glass, more familiarly known as the hour-glass. See Hour-glass; Sand-glass.

CLEPSYDRA I

Clepsydra A device for the measurement of time by the flow of running water. Its simplest form is a vessel filled with water which trickles or drops slowly from a small aperture into another vessel. One or the other of the vessels is graduated and the height of the water in that one at any given time indicates the hour. Sometimes a figure floating on the water points to the hours. Later, falling, or running, water was made to turn wheels or to move a drum, as in "Vailly's clock." Clepsydras were made and improved up to the 17th century. The earliest known example one in China is credited with having existed in 4000 B. C. The name indicates the stealing away of water and is derived from two Greek words meaning "water" and "to steal." A common form of clepsydra in India was a copper bowl with a small hole in the bottom floating on water. When the bowl filled and sank the attendant emptied it, struck the hour upon it and floated it again on the surface of the water. Like the sun-dial, the clepsydra was invented so long ago that there is no authentic record of its origin. Its evident advantages are exactly those which the sun-dial lacked. It is quite independent of day or night or other external conditions; it is conveniently made portable; and by regulating the size of the aperture through which the water flows, it can be made to work slow or fast so as, within considerable limits, to measure accurately and legibly long or short intervals of time.

The disadvantages of the clepsydra were, first, that the hole in the container tended to become worn away so as to let the water out too fast; and second, that the water ran faster from a full vessel than from one nearly empty, because of the greater pressure. This latter was in classic times corrected by a clepsydra consisting of two vessels. The second and larger of these was placed below, the water running into it, out of the first. A float within this larger vessel rose regularly as it filled, and carried a pointer which marked the time. The first vessel from which the water ran into the second, was provided with an overflow, and kept constantly full up to this level; so that the flow of water into the larger vessel remained constant.

CLEPSYDRA II

Once well established and understood in principle, the clepsydra became widely known over the ancient world, and underwent a variety of improvements and modifications in form. These latter chiefly dealt with making it more legible. Means were devised, for instance, to make it ring a bell when the water reached a certain height. And thus the alarm principle was very early brought into use. Later on, after the development of mechanical devices like the pulley and the toothed wheel or gear, the pointer was by these means constructed to move faster or slower than the rate at which the water rose, or to revolve upon a circular dial on which the hours were marked. And thus we owe to the clepsydra the origin of the modern clockface as well as of the alarm. Later still, by a more complex ingenuity, devices were arranged to strike the hours or to move mechanical figures, in fact, to perform all the functions of a clockwork which was both driven and regulated by hydraulic power. The single hour hand, however, remained in place of our two or three hands moving at different speeds, as in the modern clock or watch. The clockwork also remained primitive in construction compared with our own. Clepsydrae were always expensive, because accurate mechanical work was never cheapened until modern time. Rather they were made marvels of patient ingenuity and lavish ornament. Cunning oriental craftsmen spent their skill upon elaborate mechanism and costly decorations. The clepsydra thus became first what other time-pieces later became a triumph of the jeweler's craft a gift for kings. And the Greeks, who beautified everything that they touched, made it at once more accurate and more artistic.

The clepsydra may thus fairly claim to have been the first mechanical device for measuring time, as contrasted with the sun-dial which was really an astronomical instrument; and thus the direct ancestor of the mechanical clocks of later days. Some authorities, indeed, on the strength of certain very ancient allusions to its use in China and elsewhere, claim for it an antiquity prior to the sun-dial itself. There seems, however, to be no reason for supposing that the discovery of a mechanical law like the regular flow of water antedated so obvious a discovery as the motion of a shadow upon the ground. The explanation is probably that the invention of the clepsydra did precede the scientific perfecting of the sun-dial by the inclinations of the gnomon; which may have taken place about the time of the correction of the Babylonian calendar in 747 B. C. Not long after this date we meet with frequent references to the placing of a clepsydra in the public square of some old city, or to its use in astronomical calculations. To this, of course, its property of running by night was peculiarly adapted.

Although the chief defects of the clepsydra were minimized by the use of the two vessels and by making the aperture through which the water ran of gold or some other substance which would wear away very slowly, yet there remained certain minor imperfections. The water could not be kept entirely from evaporating; it had to be emptied out at intervals and the reservoir refilled; its accuracy was affected by the expansion of the parts under change of temperature, or it might even freeze. These faults were obviated in the sand-glass or hour-glass which for short intervals of time was also more convenient.

The clepsydra remained in use until clocks became superior to it in accuracy. See Clocks, Interesting Old; Charlemagne; Vailly.

Clerkenwell A district on the north side of the city of London within the metropolitan borough of Finsbury. It is distinguished as one of the great centers of the watchmaking and jewelers' industries in England and long established there. The Northampton Polytechnic Institute, Northampton Square, has a department devoted to instruction in all branches of the trade.

Click The click, pawl, or dog, is a necessary accessory of a ratchet wheel. It is a finger, one end of which fits into the teeth of the ratchet, while the other is pivoted on its tangent. The ratchet is thus prevented from turning backward.

Clock Specifically, a time-piece not made to be carried about but to stand upon a shelf or table, hang upon a wall or as built into a tower. Formerly the term signified particularly a time-piece which struck the hours. The word has its origin in the word for bell in Latin, gloccio; Teutonic, glocke; French, cloche; and Saxon, clugga. At one time the term was used to denote timekeepers driven by weights as distinguished from those driven by springs.

Clock-Watch A watch which strikes the hours in succession, as distinguished from repeaters. Popular in the eighteenth century.

CLOCK BANJO

Clock, Banjo-A wall clock, so called from its shape, designed by Simon Willard, of Massachusetts and very popular in its time.

CLOCK, BIRD CAGE

Clock, Bird-Cage An old form of English clock whose manufacture has been discontinued it is the oldest form of English clock still doing service. Its main feature is the endless chain drive. These clocks run thirty hours.

Clock, Bracket A form of clock very popular in England during the reign of Charles II, made to stand on a bracket or table and intended to be seen from all sides. These clocks had either a handle on top or one on each side. They were very beautifully finished.

Clock, Candle Wax or tallow candle, usually twelve inches long and marked with circular lines one inch apart. The candle would burn one inch every twenty minutes or three inches an hour. Invention credited to King Alfred the Great.

Clock, Grandfather's or Long-Case A tall clock with an anchor escapement popular thru-out the later 18th and early 19th centuries in England and America. Its excellent timekeeping qualities are due to the very long and heavy pendulum which allows a small arc of vibration. Not often made at present.

CLOCK-GRANDFATHER

Clock, Hood A style of clock originating and very popular in Holland during the late 17th century. Made of various woods, carved and ornamented and named from the hood or dome on top.

Clock, Lamp A long glass tube upright on a metal stand similar in shape to the old Roman lamps. Figures were painted on the tube to indicate the hours "12" in the middle section, with "11" above and "1" below the "12." The lamp was filled with oil up to the hour at which it was lighted then as the oil burned away the time was indicated. This form of clock was used at night in Dutch and German rural homes until a comparatively recent date.

CLOCK-LAMP

Clock, Lantern Same as Bird-Cage Clock.

Clock, Largest in World The Colgate clock in Jersey City is claimed to be twice as large as the next largest clock in the world. Its dial can be read for four miles and weighs six tons. Its minute hand is twenty feet long and the tip of it travels more than half a mile per day.

CLOCK

Clock Mysteries Glass Dial A perfectly transparent dial behind which no movement was visible. The hands were caused to revolve by watch works and semi-circular weights in the counterpoise of the hands.

Clock, Oldest in America A clock owned by the Philadelphia Public Library over two centuries old. It was made in London and is said to have been owned by Oliver Cromwell.

Clock, Sheep's-Head A clock similar to the bird-cage or lantern clock in which the dial face projects an inch or two beyond the frame.

Clock, Skeleton A clock whose works are covered with glass as a protection from dust, but are without a case, the works being exposed to view. There are eight skeleton clocks in the Charles Mifflin Hammond collection at the Essex Institute in Salem, Massachusetts.

Clock, Turret A large clock in which the dials are distinct from the movement. Because of the exposure of the hands to the wind and snow, of the clock to dust and dirt, and of the oil to freezing temperature, turret clocks to keep time must be fitted with some device to obtain a constant force on the pendulum. The first used was the remontoire but since the invention of the gravity escapement for the Westminster clock by Sir Edmund Beckett this has been used instead.

CLOCK, WAG ON THE WALL

Clock, "Wag on the Wall" A wall clock typical of the North of Holland in which weights and pendulum hung below the clock case, entirely unenclosed.

Clock and Watch Makers, English, Early For extensive lists, dates, places, and notes, see: Old Clocks and Watches and Their Makers, by Frederick J. Britten; Worshipful Company of Clockmakers, London, Published by E. J. Francis and Co., London, 1875; Old Clock Book, by Mrs. N. H. Moore.

French, Early See: Old Clocks and Watches and Their Makers, by F. J. Britten.

Scottish, Early For extensive list with dates, places and notes, see: Old Scottish Clock Makers, by John Smith.

Clock Makers, American, Early For lists, dates, places, and notes, see: Old Clock Book, by Mrs. N. H. Moore; American Clockmaking Its Early History, by Henry Terry.

Clock Mysteries; Tortoise in Water Nicholas Grollier during the first part of the eighteenth century made many mysterious timekeepers. One was a metal dish filled with water in which floated the figure of a tortoise always keeping his nose to the correct time.

Ball of Venice This was a sphere its upper and lower parts gold, and about the middle a silver band bearing the numerals. As the band revolved a Cupid's wing pointed to the hour. Its action was simple. The cord which suspended it was wound about a cylinder. The weight of the ball constituted the driving power. It had a verge escapement. The maker is not known.

Double Globe Constructed of two clear glass globes, the smaller one for the minutes above the larger hour globe. The mechanism for the latter was in the base, and for the minute globe, in the cap of the hour globe. Made by Henri Cunge.

CLOCK, OLD

Clocks, Interesting Old: Anne Boleyn's A clock said to have been presented to Anne Boleyn by Henry VIII on their wedding morning. It is about four inches square and ten inches high, of silver gilt "richly chased, engraved, and ornamented." The weights are of lead covered with copper, gilt and engraved. On one are Henry's and Anne's initials, and true lovers' knots. On the other simply H. A. At the top of each weight is "Dieu et mon droit," at the bottom "The most happye." On the top of the clock is the figure of a lion holding the arms of England, the same being engraved on the sides. The clock is now silent. There is no record as to its maker.

Canterbury This was the third of the large clocks in England. It was constructed in 1292.

Charlemagne's In 807 the King of Persia sent Charlemagne a bronze water clock inlaid with gold. The dial consisted of twelve small doors representing the hours. Each door opened at the hour it represented and the correct number of balls fell out upon a brass bell. At twelve o'clock twelve horsemen appeared and shut the doors.

Coblentz At Coblentz in a tower on the Kaufhaus is a brazen head which gnashes its teeth as the hours strike. For a Coblentzer to say "How is the man in the Kaufhaus" means "How goes it with Coblentz and the good people there?"

CLOCK, OLD

de Vick's In 1364 Henry de Vick set up a clock in the tower of the palace for Charles V. It was regulated by a balance. The teeth of the crown wheel acted upon two small levers called pallets which projected from and formed part of an upright spindle or staff on which was fixed the balance. The clock was regulated by shifting the weights placed at each end of the balance. On the bell of this clock the signal for the massacre of St. Bartholomew's was struck.

Dondi's at Pavia Built in 1344, by James Dondi, similar to Wallingford's clock.

Exeter A clock built in Exeter Cathedral sometime in the 14th century. One erected there in 1480 has the sun a fleur-de-lis which points out the hours as it revolves around a globe representing the earth. A black and white ball represents the moon's phases by turning on its axis.

Frederick II The Saladin of Egypt presented Frederick II of Germany with a clock in the year 1232. It resembled internally, a celestial globe, in which figures of the sun, moon, and other planets moved impelled by weights and wheels. There were also the twelve signs of the Zodiac which moved with the firmament.

CLOCK, OLD

Hans von Jena's An old clock in Saxony at the top of which is a very ugly head. As the clock strikes a pilgrim offers an apple on a stick to the open mouth and then withdraws it. At the same time an angel opposite the pilgrim raises her eyes from her book. The legend goes that Hans von Jena, for a crime, was condemned to undergo such torture for three centuries.

Jefferson's An old weight clock in which the weights are carried over a pulley and made to indicate the day of the week by their position. This is in the hallway at Monticello.

Lists and Descriptions of See Curiosities of Clocks and Watches, E. J. Wood. Old Clocks and Watches and their Makers, F. J. Britten. Old Clock Book, N. H. Moore.

CLOCK, VASE

Vase Clocks of Marie Antoinette The movement was inclosed in a marble pedestal. About the beautifully tinted porcelain urn was a double band, on which were marked the numerals and which revolved every twelve hours. A serpent with head erect pointed to the hour.

CLOCK, SKULL

Mary, Queen of Scots Skull Watch or Clock. A small clock in the form of a skull said to have been given by Mary, Queen of Scots, to Mary Seaton, one of her maids of honor. The skull is of silver gilt and is engraved with figures of Death, Time, Adam and Eve, and the Crucifixion. The lower part of the skull is pierced to emit the sound when it strikes, being cut in the form of emblems of the Crucifixion. The works occupy the brain's position in the skull fitting into a silver bell which fills the entire hollow of the skull. The hours are struck on this bell by a small hammer on a separate train.

Pope Sixtus' Built by Habrecht of Strasburg in 1589. It greatly resembles the Strasburg clock which Habrecht also built. It was in the possession of the Popes for more than two centuries and later became the property of William I, King of the Netherlands. In 1850 it was exhibited in England after which it became the property of Mr. O. Morgan. It performs all the feats of the Strasburg clock.

Rouen In the Rue de la Grosse Horloge in Rouen a clock made by Jehan de

Fealius in 1389 is built in a tower which surmounts an arched gateway. Its dial is about six feet square. It shows the hours, days of the week, and phases of the moon. It still keeps excellent time and is the chief clock of the city.

CLOCK, OLD

St. Dunstan's Erected in 1671 above the gateway of the old St. Dunstan's Church. The clock had two dials, back to back upheld by a quaint bracket. In a little open belfry above were the gaily painted figures of Gog and Magog which struck the quarters on bells suspended near them. In 1830 the clock was sold to the Marquis of Hertford who set it up at his home in Regent Park.

St. Paul's A clock existed prior to 1298 in the tower of St. Paul's Cathedral which struck the hours by means of mechanical figures called Paul's Jacks. Later a fine dial was added.

Strasburg Rebuilt twice after the first one which was begun about 1352. This first clock consisted of a calendar which showed the principal movable feasts. It showed also the movements of the sun and moon. On the upper part was a statue of the Virgin before which at noon the figures of the three Magi bowed. At the same time a cock automaton opened its beak, flapped its wings and crowed. 2. The second Strasburg clock was erected about 1570. This was a very elaborate mechanism, showing besides the time, a calendar for a century, the movements of the sun and moon, eclipses of the same and other things. The striking was done by an elaborate automatic arrangement. (See Old Clocks and Watches and Their Makers F. J. Britten.) 3. In 1842 the clock was again thoroughly reconstructed. This, too, is a very elaborate system of motions showing the movements of sun, moon, and planets, also sidereal time, a calendar, etc. The hours and quarters are struck by automatic figures.

Ulm In the eastern end of the old Rathaus at Ulm is installed an astronomical clock which dates from the beginning of the 16th century. It was thoroughly repaired in 1549 by the builder of the Strasburg clock Isak Habrecht. Shows in addition to the hours, the diurnal and annual revolutions of the earth and the movements and phases of the moon. The clock is an artistic achievement as well as a mechanical wonder.

Vailly's A scientific water clock. It consisted of a tin cylinder divided into several small cells and suspended by a thread fixed to its axis, in a frame on which the hour distances fixed by trial were marked. It was so made that the water passed slowly from one cell to the next and as it did so it changed the center of gravity of the cylinder and set it in motion so as to indicate the time on the frame. Made about 1690.

Wallingford's Built in 1326 in St. Alban's Monastery. It showed besides the hours, the apparent motion of the sun, the ebb and flow of tides, changes of moon, etc. It continued to run until the time of Henry VIII. Held by some to have been a mere planetarium.

Wells Cathedral Clock built by Peter Lightfoot, A. D. 1340 at Glastonbury and removed to Wells Cathedral during the Reformation, after the dissolution of the Glastonbury monastery. In 1835 it was again removed to the South Kensington museum. At that time the worn-out works were replaced by a new train, but the dial and knights were retained. The dial is divided into twenty-four hours and shows the motion of the sun and moon. On its summit are eight armed knights tilting at one another, lance at rest by a double rotary motion.

Westminster A clock said to have been erected at Westminster with the proceeds of a fine imposed upon one of the Chief Justices about 1288. About 1365 Edward III had a stone clock tower erected at Westminster. This tower contained a clock which struck the hours on a great bell. It also contained other bells. This tower was razed by the Roundhead mob about 1650. Later a dial with the motto "Discite justiam monite" was placed on the site. The bell "Great Tom" was given to St. Paul's about the beginning of the 18th Century. The present Westminster clock is made after plans by E. B. Denison (Sir Edmund Beckett) and made by E. J. Dent. The bell is called "Big Ben." It is claimed to be the best timekeeper of its kind in the world. It was for use in this clock that Denison invented his gravity escapement.

Wimborne A very old clock at Wimborne in Dorsetshire, much like the Wells Cathedral clock. By some authorities believed also to have been planned by Peter Lightfoot.

Clock-Setters During the early history of turret clocks, for each one was employed a caretaker called the "setter." That such an official was needed indicates that they were more or less undependable.

Cock A horizontal bracket. See: Balance Cock; Escape Cock; Pendulum Cock; Potance.

Collet A collar or flange on a cylindrical piece of metal. Any part of such cylinder of greater diameter than the rest. Sometimes of the same piece of metal; sometimes fitted friction tight upon it.

Compensation The provision made in a clock or watch to counteract the expansion and contraction due to variations of temperature. In the clock it is applied to the pendulum; in the watch to the balance.

Compensation balance

Compensation Balance A balance corrected for errors caused by variations in temperature. The type in most general use was invented by Thomas Earnshaw in the second half of the 18th century. The double rim of this balance is constructed of brass and steel soldered together in the form of a cut ring, the brass on the outside. When heat, elongating the balance ring, causes it to vibrate more slowly, the brass, expanding more than the steel, bends the free ends of the cut rim toward the center, thus decreasing the diameter of the balance and quickening the vibration. On the other hand, when cold, contracting the ring tends to quicken the vibration of the balance, the contraction of the brass rim draws the free end outward, making the diameter larger and the vibration slower in consequence. The compensation balance is also made with brass as the inner metal and aluminum outside.

Compensation Curb A laminated bar of brass and steel or aluminum and brass fixed at one end, the free end carrying the curb pins that regulate the length of the balance spring. Common in old watches but not now in use.

Compensation Pendulum A pendulum so constructed that the distance between the point of suspension and the center of oscillation remains constant in all temperatures. See: Pendulum, Gridiron and Pendulum, Mercurial Compensation.

contrate wheel

Contrate Wheel A wheel whose cogs are parallel to its axis and whose axis is at right angles to the axis of the wheel into which it gears. A crown wheel.

Corrosion The eating or wearing away of metals by slow degrees through chemical action.

Countersink To enlarge the outer end of a hole for the reception of the head of a screw, bolt, etc. The term is also applied to the tool with which the countersink is formed.

Coventry A municipal, county, and parliamentary borough of Warwickshire, England. One of the important watchmaking centers of Great Britain.

Crown Wheel A wheel whose teeth project at right angles to the plane of the wheel. A contrate wheel. The escape wheel of the verge escapement is an illustration.

Crutch A light rod in a clock descending from the pallet arbor and ending in a fork which embraces the pendulum rod. It transmits the motion of the pallet to the pendulum.

Ctesibus A famous Greek mechanician who lived in Alexandria about 130 B. C. Although his was not the first clepsydra as is claimed by some it was an ingenious and interesting one. Believed to have first applied toothed wheels to clepsydrae about 140 B. C.

Curb Pins See Banking Pins.

Cusin, Charles A watchmaker from Autun, Burgundy, who laid the foundation for the Swiss watch industry in Geneva in 1587. It grew very slowly at first in 1687 having only one hundred watchmakers with three hundred assistants. In 1760 there were at Geneva eight hundred watchmakers with 5,000 to 6,000 assistants.

Custer, Jacob D. (1809-1879.) A Pennsylvania clockmaker in 1831; he was one of the early makers of watches in America in 1840. However, his work was not important commercially, for he produced only about a dozen watches. A very ingenious man, who, it is said, made everything from a steam engine to his own shoes. He made hundreds of the clock movements which at that period were used to revolve the lanterns in lighthouses.

Cycle of the Sun A period of twenty-eight years, after which the days of the week again fall on the same days of the month as during the first year of the former cycle. It has no relation to the sun's course but was invented for the purpose of finding out the days of the month on which the Sundays fall during each year of the cycle. Cycles of the sun date from nine years before the Christian era.

Cycloid A curve generated by a given point in the circumference of a circle which is rolled along a straight line always in the same place. Example: The curve traced by any point in the rim of a wheel which travels in a straight line along a level road.

Cylinder Escapement See: Escapement, Cylinder.

Cylinder Plugs Plugs fitted into the ends of the cylinder of a cylinder escapement. Their outer extremities are formed into the pivots on which the cylinder rotates.

Damaskeen To decorate a metal by inlaying other metals or jewels, or by etching designs upon its surface. To be distinguished from snailing, with which it is often confounded.

damaskeen

Day The time of one complete revolution of the earth on its axis. The actual length of this day is continually changing owing to the eccentricity of the earth's orbit and the angle of the ecliptic. The mean solar day is 24 hours. The sidereal day is 23 hours, 56 minutes, 4.099 seconds.

Day, Nautical The nautical day begins when the sun is on the meridian and eight bells are struck. The day is divided into "afternoon watch" or four hours, two "dog watches" of two hours each, then "middle watch," "night watch," "morning watch" and "forenoon watch," each of four hours, completing the day.

Denison, Edmund Beckett Sir Edmund Beckett Lord Grimthorpe. Born 1816. A lawyer by profession, and the inventor of the gravity escapement for turret clocks; also an authoritative writer on horological subjects. He designed and planned the Westminster clock said to be the best timekeeper of its kind in the world. Died 1905.

Dennison, Aaron L. Born in Freeport, Me., in 1812. Died Birmingham, England, January 9, 1898. At eighteen he was apprenticed to a watchmaker. Later in working at the trade, he was impressed with the inaccuracies which existed in the best handmade watches. This, with a visit to the Springfield Armory, gave him his idea of machine-made watches with interchangeable parts. He interested Edward Howard in the project, and having found the needed capital they started in the business and laid the foundation of what is now the Waltham Watch Company. Dennison has been called the "father of American Watchmaking" tho there seems ground for the claim that he shares that honor with Edward Howard.

Depthing The technical name for the proper adjusting or spacing of the gearing in a watch.

Detent The device which halts, and releases, at the proper instant the escapement of a clock or chronometer. See: Escapement.

de Vick, de Wyck, or de Wieck, Henry A German clockmaker who, in 1364, made the first turret clock of which reliable information and description remains. The clock was made for Charles V. See: Clocks, Interesting Old De Vick's.

Dial Commonly called the face of the watch made of gold or silver or other metal or of enamel, with the required figures in the United States one to twelve upon it in a contrasting color. See also, Sun-dial.

Dial Feet Short wires soldered to the back of the dial of a watch or clock which hold it in place by fitting into holes in the pillar plate.

Dial of Ahaz A sun-dial belonging to Ahaz, King of Judea 742-727 B. C., mention of which occurs twice in the Scriptures II Kings, XX: 9-11, and Isaiah XXXVIII: 8. It is believed that one of his Babylonian astrologers constructed it for him.

Dial Plate See Lower Plate.

Dial, Sun See Sun-dial.

Dial Wheels The wheels constituting the motion work of a watch.

Diurnal In an astronomical sense, pertaining to a period covering a mean solar day. See: Solar Time.

Dog Screw A screw with an eccentric head used to attach a watch movement to a dome case.

Dog-Watch A nautical term for two daily two-hour periods of watching aboard ship. The first begins at 4 P. M., the other at 6 P. M.

Dolmen A sacred instrument used for astronomical purposes at certain critical periods of the year; formed of four stones at the cardinal points and a leaning stone crossing diagonally and forming with the east stone a sacred "creep-way." The solar hours were indicated by the shadow of the leaning stone touching various prominent points or edges. One at Camp, England, is prehistoric.

Dome The inner case of a watch which snaps on the band of a case.

Dome-Case A case in which the inner case or dome snaps to the band of the case.

Dondi, Giacomo Born at Padua, Italy, in 1298. In 1344 he set up at Padua a famous clock which became a model for later clocks and which earned for him the surname, "Orologio."

Double Bottom Case A watch case in which the inner cover or bottom is made solid with the middle. The vogue in English cases for a long time;

now almost obsolete.

Double-Sunk Dial A dial in which there are two sinks; one for the hour hand, and a deeper one for the seconds hand.

Draw 1. The force which holds the lever against its bank, due chiefly to the angle of the locking face of the pallet stone. 2. The angle of the locking faces of pallets in the lever escapement.

Driver Of two wheels working together, the one which imparts the power. The driven wheel is termed the follower.

Driving Wheel In a clock the wheel on the main arbor which drives the whole train.

Drop That part of the motion of the escape wheel when it is not in contact with the pallet.

Drum The cylinder, or barrel, on the main arbor in a clock on which the driving cord winds, raising the weight, when the clock is being wound.

Dummy Watch (Fausse Montre.) About 1770 it became the fashion to wear two watches. But because two real watches were too expensive for most people, the custom grew up for having one sham watch usually worn on the right side. These were called "dummy watches" or "fausse montres."

Earnshaw, Thomas 1749-1829. An eminent English watchmaker who invented the spring detent escapement and the compensation balance, both essentially the same as are now used in chronometers. He first soldered brass and steel together for the balance instead of riveting them.

East, Edward Watchmaker to Charles I and an eminent horologist. He was one of the ten original assistants named in the charter of the Clockmakers' Company and at once took a leading part in their proceedings. He was elected master in 1664 and 1682. He was the only treasurer ever appointed by that company. He died probably about 1693. East's watches were often presented as prizes by Charles in tennis tournaments.

Edward VI King of England from 1546 to 1553. Said to have been the first Englishman to wear a watch.

Electric Clock A clock in which the pallets moved electrically from a distant mechanism drive the escape wheel and the hands.

Ecliptic That plane passing through the center of the sun in which lies the orbit of the earth. Also used to designate the apparent path of the sun in the heavens.

Elgin A city in Illinois, U. S. A., in which is located the Elgin National Watch Company one of the largest factories in the United States.

End-Shake Freedom of pivots to move endways. Necessary in a watch or clock because there is no force to spare and a tight pivot would stop the movement.

End-Stone A small disc of jewel against which the end of a pivot sets. See Capped Jewel.

End-Stop In a watch the same as end-stone.

Engaging Friction Friction which results when the teeth of two wheels gearing together come into action before reaching the line of centers that is, a line drawn from center to center of the gearing wheels.

Engine-Turning A pattern of curved lines cut into metal for decoration. Introduced about 1770 by Francis Guerint of Geneva. The earliest specimens were cut very deep but shallower cutting soon became the rule.

Engraving A form of ornamenting metals in which the design is cut into the metal. In "Champ-leve" engraving the ground is cut away leaving the design in relief.

Epact The excess in time of the solar year over the period of 12 lunar months, amounting to about 11 days. The new moons will thus fall about 11 days earlier in each succeeding year. In a calendar so arranged 30 days are taken off every fourth year, as an intercalary month, the moon having revolved once in that time, and the three days remaining would be the epact. The epact thus continues to vary until at the end of nineteen years the new moons return as at first.

Epicycloid A curve generated by any point in the circumference of a circle as it rolls on the outside of the circumference of a fixed circle. This curve is the best for the face of the teeth of a driving wheel.

Equation Clocks An obsolete form of clock which showed true solar or sun-dial time instead of mean solar, or average time.

Equation of Time The difference between true time and mean, or averaged time. There are four days in the Gregorian year when the true time and mean time agree, and the equation of time is zero: These are December 24, April 15, June 15, and August 31. Between the first two dates and the last two dates, true time is earlier than mean time; for the other two periods of the year it is later.

Escape Cock The bracket which supports the upper ends of the escape wheel and pallet staff arbors.

Escapement The device in a watch or clock which regulates the motion of the train thus distributing the power of the main-spring. It communicates the motive power to the balance or pendulum. Escapements are of three classes: recoil, dead, or dead-beat; and detached.

escapement

Escapement, Anchor The recoil escapement, invented by Hooke, used in most house clocks. A name also applied to one kind of Lever Escapement with an unusually wide impulse pin. The recoil escapement is one in which each tooth of the escape wheel, after it comes to rest, is moved backward by the pallets. Altho one of the easiest escapements to set out correctly the pallets are often improperly formed making an escapement which gives indifferent service. As a timekeeper the anchor escapement is inferior to the dead-beat escapement.

Escapement

Escapement, Chronometer A detached escapement in which the escape wheel is locked on a stone carried in a detent, and in which the teeth of the escape wheel impart an impulse to a pallet on the balance staff with every alternate vibration. Used in Marine Chronometers.

Escapement, Crown-Wheel Of the recoil type, and the earliest known escapement; to be found in Henry de Wyck's clock. Not suitable for watches. Practically the same principle as Verge or Vertical Escapement used in watches for so many years.

Escapement

Escapement, Cylinder or Horizontal Invented by Thomas Tompion in 1695 later improved and brought into general use by Graham. It dispensed with the then common vertical crown-wheel hence the term "horizontal" and permitted thinner watches. This escapement is frictional, the balance being carried on a hollow cylinder whose bore is large enough to admit the teeth of the escape wheel. The cylinder is cut away where the teeth enter and the impulse is given by the wedge shaped teeth striking against the edge of the cylinder as they enter and leave. Used at this time in the cheaper Swiss watches.

Escapement Escapement

Escapement, Dead-Beat Any escapement in which the pallet face is so formed that the escape wheel remains dead or motionless during the supplementary arc of the balance or swing of the pendulum. As invented by George Graham, the wheel is much the same as the wheel in the anchor escapement, the difference lying in the shape of the pallets. Each pallet has a driving face and a sliding face. It is so arranged that the impulse is given the pendulum at the midpoint of its swing thus allowing the swing to adapt itself to the impulse and keep the time constant. The pallets are faced with jewels so that there is slight friction. Used in high grade clocks such as regulators and astronomical clocks.

Escapement, Detached Any escapement in which the balance or pendulum is for some time during each vibration free from the pressure of the train. Detached escapements are used in chronometers, most watches and in turret clocks. They are of value in any movement where the motive power varies greatly hence in turret clocks. Examples: Chronometer, lever, and gravity escapements.

Double escapement

Escapement, Double Three-Legged Gravity Invented in 1854 by E. B. Denison, Esq., for the great clock at the Houses of Parliament. It is the best escapement for very large clocks where the hands are exposed to the action of the wind and snow, because it admits of great driving power in the movement without its sensibly affecting the escapement as would be the case in the dead-beat type. The impulse to the pendulum is given by the weight of the lever arms falling through a given distance and is therefore constant. This escapement consists of two gravity impulse pallets pivoted in a line with the bending point of the pendulum. There is a locking wheel made of two thin plates of three teeth each. Between these plates are the three pins that lift the pallets. The locking is effected by blocks screwed to the front of one pallet and the back of the other. Impulse is given by the pallets in turn striking the pendulum rod. The pendulum rod serves to unlock the wheel. The arrangement is such that the lifting pins have a little free run each time. Since the pallets are always lifted the same distance they give a constant impulse to the pendulum.

Duplux escapement

Escapement, Duplex Invented by Hook; later improved by Tyrer. Very accurate but as originally made was affected by any sudden motion, and hence of little use in watches. The escape wheel has two sets of teeth. Those farthest from the center lock the wheel by pressing on a hollow ruby cylinder fitted round the balance staff and notched so as to permit the passing of the teeth as the balance moves in a direction opposite to the wheel's motion. The second set stand up from the face of the wheel and one gives impulse to the pallet every time a tooth leaves the notch. This is not a detached escapement, but there is little friction. As improved this escapement was used in the famous Waterbury watches.

Escapement, Foliot A form of escapement actuated by a foliot balance. See Foliot.

Four-legged gravity escapement

Escapement, Four-Legged Gravity Invented by E. B. Denison (Sir Edmund Beckett). The same in principle as the Double Three-Legged escapement, only it has but one escape wheel with four teeth or legs instead of two wheels with three legs each. The wheel has two sets of lifting pins one acting on each pallet. Occasionally used in regulators and other clocks with a seconds pendulum, but of doubtful, if any, advantage over the Graham dead-beat escapement.

Escapement, Frictional Any escapement in which the balance is never free from the escapement. Examples: The Cylinder, Duplex and Verge types.

Escapement, Gravity An escapement which gives impulse to the pendulum by means of a weight falling through a constant distance. Of use in turret and other exposed clocks where the hands' movements are affected by wind, rain, and snow. See subtitles under these headings: Double Three-legged Gravity; Single Three-legged Gravity; Four-legged Gravity; Six-legged Gravity.

Escapement lever

Escapement Lever Invented by Thomas Mudge about 1765. It is the preferred escapement for watches because of the certainty of its performance. Possibly inferior to the chronometer escapement as a timekeeper. Its most noticeable defect is the necessity of applying oil to the pallets, the thickening of which affects the action. There are many other kinds of lever escapements. The Mudge escapement was essentially like the modern Double Roller. The connection between the balance and the escape wheel is made by a lever to which the pallets are fastened, and into the forked end of which plays the ruby pin which is carried on a roller on the same staff as the balance. Each pallet has an impulse face and a locking face. The impulse is given by the escape wheel tooth striking the impulse face of a pallet and communicated to the balance by the lever, raised by the pallet's movement striking the ruby pin in the roller. This ruby pin also serves to unlock the pallets by causing the lever to lift them in turn. This escapement is of the detached type. The action of the lever is kept within the desired limits by banking pins.

Escapement, Lever Club Tooth An escapement like the Table Roller in the action of the lever and roller, but differs in the pallet action. The impulse planes are partly on the teeth and partly on the pallet. This is the standard watch escapement of today.

Escapement, Crank Lever An escapement with a small roller having a tooth like a pinion leaf projecting from its circumference. This tooth acts in a square notch cut in the end of the lever. The lever is formed like a fork the two points of which act as safety pins against the edge of the roller to prevent the lever from getting out of action with the roller. It necessitated very careful construction and was not so good as the Double Roller or Table Roller.

Lever-Double roller escapement

Escapement, Lever Double Roller This escapement has two rollers on the balance staff, the large one carrying the balance staff and the small one used for a safety roller only. The best form of lever escapement but more delicate, expensive, and difficult to make than the Table Roller; hence not so much used as the latter.

Escapement, Patent Detached Lever Introduced in 1766 by Thomas Mudge, but neglected for years thereafter even by Mudge himself. It was in some of its parts the model of the best form of lever escapement the Double Roller. The first pallets had no "draw" on the locking faces which rendered the escapement peculiarly sensitive to jolt and jar. This may have suggested to Mudge the addition of the small roller, whose worth has been since unquestionably demonstrated.

Escapement, Lever Pin-Pallet A lever escapement with round pins for pallets, and the inclines on the escape teeth. Used in alarm clocks.

Escapement, Rack-Lever Invented by Abbe Hautefeuille in 1734. Afterward made and improved by Berthoud and by Peter Litherland, who obtained a patent for it in 1794. It consisted of anchor shaped pallets on whose axis was fixed a rack, or segment of a toothed wheel which geared into a pinion on the axis of the balance. The balance was thus never free from the train and good timekeeping was made impossible. It is not now in use.

Lever-resilient escapement

Escapement, Lever-Resilient Invented by F. J. Cole about 1870. A form of lever escapement designed to obviate the evils of overbanking. The points of the escape-wheel teeth are bent toward the locking faces of the pallets, the bend in the tooth acts as the banking and no pins are required. It was abandoned because expensive to make and the danger of overbanking is not considerable.

Escapement, Lever Table Roller Excellent and very simple and the most common form today. It differs from the crank lever only in the action of the roller. The impulse pin instead of projecting beyond the edge of the roller is set within its circumference and raised above its plane.

Escapement, Lever Two Pin A form of Lever Escapement in which the unlocking and impulse actions were formerly divided between two small gold pins in the roller and one in the lever. Later the two roller pins were discarded, and one broad jewel pin substituted.

Escapement--pin wheel

Escapement Pin Wheel Invented by Lepaute about 1750. Similar in action to the dead-beat. A good and simple escapement for large clocks. The impulse is given the pendulum through the pallets by pins which stand out from the face of the escape wheel. Lepaute made these pins semi-circular and had his pallets of equal length acting on opposite sides of the wheel. Sir E. Beckett cut away part of the front of the pins which allows the pallets to act as in the diagram. The resting faces are arcs of a circle. It has been superseded by the gravity escapement for large clocks and is inferior to the dead-beat for small.

recoil escapement

Escapement, Recoil Any escapement in which the pallets actually force the escape wheel to turn backwards a trifle with each beat of the balance. Cheap and easy to make but inferior as timekeepers to the detached or dead-beat types.

Escapement, Right-Angled A lever escapement so set that lines drawn between the centers of the balance, pallets, and escape wheel would form a right angle. See Escapement, straight-line.

Escapement, Single-Beat An escapement such as the Duplex, or Chronometer, whose escape wheel moves only at alternate beats of the balance or pendulum.

single three legged gravity escapement

Escapement, Single Three-Legged Gravity Consists of two pallets and one three-legged locking wheel. Instead of the three pins for lifting as in the Double Three-Legged Gravity escapement there is a triangular steel block which acts against large friction rollers, pivoted one on each pallet.

Escapement, Six-Legged Gravity A modification of the three-legged gravity escapement. The locking wheel has six teeth. One of the pallet arms is neutral and gives no impulse, hence impulse is given only at each alternate vibration. A much lighter driving weight than for the Double Three-legged Gravity escapement will suffice for this, since the rotations of the escape wheel required are only half as many.

Escapement, Straight-Line An escapement of the lever type in which the escape wheel, pallets and balance are all in a straight line; an arrangement favored by the Swiss.

Escapement, Verge Also called "Crown-wheel," or Vertical escapement. The earliest form of escapement on record. The inventor is not known, but the escapement was used on de Vick's clock. (1364.) It was used almost exclusively up to 1750 in spite of its manifest inaccuracy. The verge is a frictional recoil escapement. It consists of a crown-wheel, with eleven, thirteen, or fifteen teeth, shaped like those of a rip saw, and with its axis set at right angles to the pallets axis, or verge, which carries the balance. The verge is a slender cylinder as small as compatible with the required strength, from which project the pallets, two flat steel "flags" at an angle to each other varying from 90° to 115°. The wheel runs in a watch in a plane at right angles to the face. Any variation in the motive power causes a variation in the arc of the balance swing. Therefore, since the time of oscillation depends on the arc of the swing, the time-keeping qualities were directly affected. This gave rise to the invention of the stack-freed and fusee, both contrivances to equalize the power of the mainspring. In spite of the many defects the verge escapement was one of the great inventions because the first escapement, and was used for centuries before superior kinds were devised. It necessitated thick and bulky watches.

Escapement, Virgule An early form of escapement invented about 1660 by Abbe Hautefeuille. Its action can be readily understood from the diagram.

Escape Pinion The pinion on the escape-wheel arbor.

Escape Wheel The last wheel of a train: it gives impulse to the balance, indirectly. Also called scape wheel. Easily identified by teeth resembling those of a circular saw.

Face 1. Of a watch or clock is the dial. 2. Of the tooth of a wheel, that portion beyond the pitch line.

Facio, Nicolas A Geneva watchmaker who invented the art of piercing jewels for use in watches, and in May, 1705, obtained a patent therefor in London. In December of the same year when he petitioned for a more extended patent he was opposed by the Clockmakers' Company, who produced in evidence proof that Facio was not first in this use of jewels, in an old watch of Ignatius Huggeford's with an amethyst mounted on the cock of the balance wheel. Facio's petition was denied. It was later discovered that Huggeford's jewel had nothing to do with the mechanism of the watch.

Favre, Perret E. In 1876 the chief commissioner in the Swiss Department and a member at that time of the International Jury on Watches at the Centennial Exhibition at Philadelphia. On his return home he was very emphatic in his endorsement of the American method of manufacture as compared to the Swiss.

Fitch E. C. Made president of the Waltham Watch Co., in 1886. His long experience in watch case and movement making and his commercial training made his judgment on matters relating to watchmaking of value. He was the inventor of the screw bezel case.

Flank The flank of a wheel or pinion is the part lying between the pitch circle and the center.

Flirt Any device for causing the sudden movement of a mechanism.

Fly A speed regulating device or governor consisting of a fan or two vanes upon a rotating shaft. Used in the striking part of clocks. By some believed to have been used on the earliest clocks before the verge escapement to check a too rapid descent of the weight.

Fly Pinion The pinion in a clock that carries the fly: a part of the striking mechanism.

fob

Fob Properly a watch pocket in the waistband of trousers. Commonly applied to the end of a chain or ribbon which is attached to the watch and hangs free from the pocket. One of the early examples was attached to a watch made for Oliver Cromwell in 1625 by John Midwall in Fleet Street.

foliot

Foliot A straight armed balance with weights used as one of the earliest clock regulators. De Vick's clock is one example of it.

Foliot Balance See Foliot.

Follower Of two wheels geared together, the one to which the driver imparts motion is called the follower.

Fork The fork shaped end of the lever into which plays the roller jewel.

Fourth Wheel The wheel in a watch that drives the escape pinion and to whose arbor the seconds hand is attached.

Frame The plates or plate and bars of a watch or clock which support the pivots of the train.

Free Spring A balance spring not controlled by curb pins. Used in chronometers and other fine time pieces where the spring is an overcoil.

Fromanteel, Ahasuerus A clockmaker of Dutch extraction maker of steeple clocks in East Smithfield. The family of Fromanteels were celebrated as having been the first to introduce the pendulum clocks into England. Their claim has since been contested in favor of Harris and Hooke.

Full Plate A model in which the top plate is circular in form the balance being above this plate. Used now in 18 size watches for railroad and other hard usage. They are made only in limited quantities.

fusee

Fusee Invented by Jacob Zech of Prague about 1525. Consists of a specially grooved cone-shaped pulley interposed between the mainspring barrel and the great or driving wheel of a watch or clock. The connection between the barrel and fusee was first made by a cord or catgut, later by a chain. In winding the spring the cord is drawn from the barrel on to the fusee the first coil on the larger end. Thus the mainspring when fully wound uncoils the cord first from the smaller end of the fusee; and as it runs down gets the benefit of increased leverage by reason of the greater diameter of the lower part of the fusee. An excellent adjustment of the pressure on the center pinion can be made in this way. The fusee has been abandoned in watches to allow of thinness, but is still used in chronometers and clocks.

Fusee Cap A thin steel plate with a projecting nose on the smaller end of the fusee: a part of the mechanism to stop the fusee when the last coil of the chain is wound thereon.

Fusee Chain A very delicate steel chain connecting the barrel with the fusee of a watch, chronometer or clock. It replaced the catgut originally used and was first introduced by Gruet of Geneva about 1664.

Fusee Sink The sink cut in the top plate of a watch to give space for the fusee.

Galileo, Galilei Commonly called "Galileo." A famous Italian scientist born in 1564 who discovered, among many other things, the isochronism of the pendulum vibrating through long or short arcs. The story goes that he noticed that a swinging chandelier in a certain cathedral took the same length of time to each vibration whether in long or short arcs timing them by his pulse. He seems never to have applied this principle to clocks, although he issued an essay on the subject in 1639.

Galileo, Vincentis Son of the great astronomer, born about 1600. He aided his father in experiments and gave special attention to the application of the pendulum to clocks. He is claimed by some to have been the first to so apply the pendulum, in 1649, but this is disputed in favor of Richard Harris of London.

Geneva A city in Switzerland in which watchmaking was first established in that country. It is the center of the "hand" industry, and the city is honeycombed with garret-workers so-called making parts.

Gerbert (Pope Sylvester II) Born in Belliac, Auvergne, in 920. In 990 Gerbert made some sort of a clock which attained wide fame. Some authorities claim that it was a clock moved by weights and wheels and some even claim for it a verge escapement. On the other hand, other authorities state positively that that story is a myth and that Gerbert's horologe was a sun-dial. It seems pretty well accepted that there was no escapement used, however, until more than two centuries after Gerbert's time.

German Silver An alloy of copper, nickel, and zinc copper predominating.

Really a white brass.

Gimbal A contrivance resembling a universal joint permitting a suspended object to tip freely in all directions. Marine chronometers are supported in their cases or boxes by gimbals. It was first applied to chronometers by Huyghens.

gnomon

Gnomon A simple and probably the most ancient instrument for marking time consisting simply of a staff or pillow fixed perpendicularly in a sunny place time being reckoned by the changing length of the shadow or by its angular movement. In more recent times the title "gnomon" was applied to the style of the sun-dial.

Gnomonics The art of constructing and setting sun-dials taught especially in the seventeenth century.

Goddard, Luther Born at Shrewsbury, Mass., February 28, 1762 Died 1842. He was the first American to manufacture watches. He began in 1809 but unable to compete as to price with cheap foreign watches, retired after making about five hundred.

going barrel

Going-Barrel The Swiss early abandoned the fusee in watches and cut teeth around the outside of the main-spring barrel so as to drive the train direct. Such an arrangement is called a going-barrel. It made possible a thinner and much simpler watch. American makers quickly adopted this device, but the English long clung to the fusee. It is sometimes claimed that the French were the first to adopt the going-barrel.

Going Fusee A fusee with maintaining power attachment, so that the watch does not stop while being wound. Invented by Harrison.

Golden Number Meton, an Athenian astronomer, discovered about 432 B. C. that every nineteen years the new and full moons returned on the same days of the month. This period is the cycle of the moon, called the Golden Number because the Greeks, to honor it, had it written in letters of gold. Anno Domini, the year of our Lord, fell on the second year of a lunar cycle. Hence, to find the Golden Number for any year, add 1 to the date (A. D.) and divide by 19. The remainder is the Golden Number for the year.

Gold-Filled A sheet of brass sandwiched between two thin plates of gold and all brazed together. Gold-filled watch cases were introduced in America. They give very good wear.

Graham, George, F. R. S. An English watchmaker and astronomer, born in Cumberland in 1675. Died 1751. He was an apprentice of Tompion and succeeded to Tompion's reputation as the best watchmaker of his time. He invented the mercurial compensation pendulum, the dead-beat escapement, and perfected the cylinder escapement of Tompion and left it in practically its present form. He made ornamentation distinctly subsidiary to use. He was master of the Clockmakers' Company in 1722-23. He was buried with

Tompion in Westminster Abbey.

Great Tom The great bell which struck the hours on the first clock at Westminster. It was afterwards transferred to St. Paul's.

Great Wheel In a fusee watch the toothed wheel which transmits the power from the fusee to the center pinion. In a going-barrel watch it is represented by the toothed portion of the barrel drum.

Greenwich Observatory (England) Royal observatory founded 1675 to promote astronomy and navigation. There is at this observatory a standard motor clock which is the center of a system of electrically controlled clocks scattered over the Kingdom, and which thus keeps official time as our Naval Observatory clock does for the United States.

Grimthorpe See Denison, E. B.

Gruen, Dietrich A Swiss watchmaker who with his son Fred first succeeded in making a very thin watch. The Gruen watch factory at Cincinnati, Ohio, is unique in this country. The buildings and surroundings resemble those of Switzerland, and the method of manufacture embodies more handwork than is common in the American system.

Gruet A Swiss who introduced chains for the fusee instead of catgut cord, in 1664. They are still used for marine chronometers, some clocks, and the few fusee watches now made.

Guard Pin A pin in a lever escapement which prevents the pallets leaving the escape wheel when the hands of a watch are turned back. Also known as the "safety pin."

Guild or Gild An association of people occupied in kindred pursuits for mutual protection and aid. Watch and clockmakers belonged to the Blacksmiths' Guild in England until 1631, when the Clockmakers' Company was formed. In France the Clockmakers' Guild was powerful in 1544.

Hair-Spring Said by some to be a distinctly American term for the balance spring of a watch. But Wood (English) uses it in his "Curiosities of Clocks and Watches," 1866. However, it is not in common use outside of America. It is thought to have originated from the fact that in early times attempts were made to utilize hog-bristle for the balance spring.

Half Plate A watch in which the top plate covers but half of the pillar plate, the fourth wheel pinion being carried in a cock to allow the use of a larger balance. Now obsolete or nearly so. Replaced by the bridge-model.

Hall Mark A stamp placed upon gold and silver articles by government officials after the metal therein has been assayed.

Hands The metal pointers which, moved by the train, indicate the time by pointing to the figures on the dial. At present there are always two, the hour and minute hands and frequently a seconds hand also. Clocks at first were made with only the hour hand; the minute hand was introduced when the use of the pendulum made timekeeping sufficiently accurate for the indication of such small divisions.

Hanging Barrel A going-barrel with its arbor supported only at the upper end.

Harris, Richard An English clockmaker for whom it is claimed that he made the first pendulum clock set up at St. Paul's, Covent Garden, in 1641. Most authorities agree, however, that this honor belongs to Huyghens.

Harrison, John An English mechanician born at Faulby in Yorkshire in 1693. He made many improvements in the mechanism of clocks, the greatest of which was the compound pendulum. He won in 1761 a reward offered by Parliament in 1714 for an instrument that would determine longitude within thirty marine miles. Harrison's chronometer gave it within eighteen miles. He invented the going fusee, the gridiron compensation pendulum and suggested the idea for the compensation balance, afterward worked out by other watchmakers. Died 1776.

Hautefeuille, John (Abbe.) Born 1647. Died 1724. He disputed successfully Huyghens' claim to a prior invention of the steel balance spring. He is also credited with the invention about 1722 of the rack-lever escapement.

heart-piece

Heart-Piece The heart-shaped cam on the center-seconds wheel of a chronograph, which causes the hand to fly back to zero.

Hele, Peter (See Henlein, Peter.) Some historians credit invention of first watch to Peter Hele. There is no doubt, however, that Hele and Henlein were one and the same. Preponderance of authority favors "Henlein" as the correct spelling of the name.

Helical Following the course of a helix or spiral.

Heliotropion See "Polos."

Hemicycle Form of sun-dial in which the shadow of a vertical pointer or "gnomon" is cast upon and moves around the inner surface of a half globe or sphere. Supposed to have been invented about 350 B. C. (See Sun-Dial). Vitruvius, the Roman Engineer, ascribes invention to the Babylonian priest and astronomer, Berosus.

Henlein, Peter Sometimes called Peter Hele. A clockmaker of Nuremberg, who is believed to have made the first portable (pocket) clock or watch sometime early in the sixteenth century. Born 1480. Died about 1540. His clock was round, driven by a spring and had small wheels of steel. It was much larger than present day watches.

Hollow Pinion A pinion bored through the center. The center pinion in many watches is hollow.

"Hon-Woo-Et-Low" or Copper Jars Dropping Water A form of clepsydra at Canton, China, said to be between 3000 and 4000 years old. It consists of four copper jars arranged on steps. Each jar drops water into the one below it until the last one, in which a bamboo float, indicates the time in a rude way.

Hooke, Robert, M. D. An English physician-philosopher born on the Isle of Wight in 1635. His accomplishments were numerous. He claimed to have discovered the isochronism of the balance spring and its application to watches, though this was also claimed by Huyghens. He invented a pendulum timekeeper for finding the longitude at sea; devised the first wheel-cutting engine about 1670; and he invented the anchor escapement for clocks. His studies and inventions covered a wide field. He died in 1702.

Horologe, (Orologe), (Horologium) A general term applied indiscriminately in old writings to any mechanism for measuring time.

Horological Institute British An association of watchmakers founded in 1858 for the purpose of advancing the horological arts.

Horological Periodicals, American American Jeweler, (Monthly), Chicago, Ill.; Goldsmith and Silversmith, (Monthly), New Haven, Conn.; Jeweler's Circular, (Weekly), New York,; Keystone (Monthly), Philadelphia, Pa.; Manufacturing Jeweler, Providence, R. I.; Mid-Continent Jeweler, Kansas City, Mo.; National Jeweler, (Monthly), Chicago, Ill.; Northwestern Jeweler, St. Paul, Minn.; Pacific Goldsmith, (Monthly), San Francisco, Cal.; Trader and Canadian Jeweler, Toronto, Canada.

Horologium See Horologe.

Horology-The science of time-measurement or of the construction of time pieces.

Hour Now consisting of sixty minutes or one twenty-fourth of an equinoctial day. Formerly one twelfth of the time between sunrise and sunset, and one twelfth of the time between sunset and sunrise; hence of different lengths for day and night in the different seasons. This required much adjustment of clocks; and automatic devices for such adjustment were in great demand. A standard hour of uniform length for all times and seasons was not adopted in Paris the last place to change until 1816.

hour-glass

Hour-Glass A device for measuring hours. It has two cone-shaped superimposed glass globes connected at their apexes through a small opening. The glass contains just that quantity of sand, or mercury, as will flow in one hour through the opening from the upper globe to the lower. When it has run through the glass is reversed. See: Sand Glass. Like the sun-dial and the clepsydra, the hour-glass is older than we know. Its use probably followed close upon that of the clepsydra, or may even have preceded it in dry countries like Egypt and Babylonia, where sand was all about and water was not a thing to waste. Of its original forms there is no authentic record. Dry sand does not, like water, run faster or slower through a given opening according to the pressure from above; its rate is the same whether the upper glass is full or nearly empty. Also the hour-glass never needs to be refilled, but only to be reversed, and the same sand used over and over again. On the other hand, its convenience diminished as its size increased. It was too clumsy for use if made large enough to run

187

without attention for more than an hour or two; and in so large a glass there was more danger that the sand, however dry, might cake up and stop running. It must somehow have been transparent for convenient reading, because sand can register the time only by its flow: it cannot be made to raise a float or work a pointer. But the Egyptians very early learned to manufacture glass, and there were other substances. A legend ascribes the invention of the sand-glass to Luitprand, a Carthusian monk of the Eighth Century A. D. But this, if there is any truth in the story at all, must have been some improvement or reinvention after the forgetfulness of the Dark Ages. The device is plainly shown in Greek sculptures antedating the Christian era. Nowadays the sand-glass has pretty much disappeared, except as a kitchen timepiece for boiling eggs and the like.

Hour Hand The hand of a watch or clock which indicates the hour: for long after clocks were first made, the only hand provided.

Hour Wheel The wheel which revolves on the minute wheel or cannon pinion and carries the hour hand.

Howard, Edward Born at Hingham, Mass., October 6, 1813. Having served a regular apprenticeship in clockmaking he entered into partnership with D. P. Davis, at the age of 29, to make clocks. He was a clever mechanic and invented many pieces of mechanism, among them the swing rest. In 1849 he and Davis with A. L. Dennison and others organized the American Horologe Company for the manufacture of watches by machinery, and with the parts interchangeable the American principle of today. Though they were not financially successful the American watch industry owes its present day success largely to this beginning by Edward Howard and Aaron L. Dennison. The first company developed into the present Waltham Company, and later Mr. Howard established the E. Howard Co., at Roxbury, but severed his connection with them in 1882 and retired from business. He died March 5, 1904.

Huggeford, Ignatius An English watchmaker, one of whose watches was used to defraud Facio of his patent on the use of jewels in watches. See Facio, Nicolas.

Hunter, or Hunting-Case A watch case which has a solid metal cover over the dial.

Hunter, George Identified with watchmaking in America since about 1860 in the Waltham and Elgin Companies. He was general superintendent of the latter from 1872 to 1903, after which he was made consulting superintendent.

Huyghens, Christian A celebrated Dutch astronomer and mathematician born at The Hague, April 14, 1629. Although the honor is claimed for Richard Harris in 1641 and for Vincent Galileo in 1649 it seems historically established that Huyghens in 1657 was the first to apply to clocks the theory of the isochronism of the pendulum which the great Galileo had discovered. In 1669 he published his important work, "Horologium Oscillatorium." In 1673 he made the first clock with concentric hour and minute hands. He died in 1695.

Huyghens' Checks The arc of a swinging pendulum is a segment of a circle. For perfect isochronism it should be a cycloidal segment. To accomplish this Huyghens fixed curved brass pieces called checks for the cord to strike against but he caused thereby a greater error than he remedied. This end was later accomplished by suspending the pendulum by means of a flat steel strip instead of a cord; a device credited to Robert Hooke.

hypocycloid

Hypocycloid A curve generated by any point in the circumference of a circle which is rolled on the inner side of the circumference of a larger fixed circle.

idler

Idler, Idle Wheel, or Intermediate Wheel A toothed wheel used to connect driver and follower wheels so that both shall rotate in the same direction.

Impulse The push transmitted to the pallet by the escape wheel.

Impulse Pin The jewel pin usually a ruby on the table roller of the lever escapement, which playing into the fork of the lever transmits the impulse to the balance.

Independent Center-Seconds A watch peculiarly adapted to the use of the medical profession. It carries on a separate train a long seconds hand in addition to the hands of the ordinary watch which can be stopped without stopping the watch.

Independent Seconds A watch whose seconds hand is driven by a separate train.

Ingersoll, Charles Henry Secretary, Treasurer and General Manager of Robt. H. Ingersoll and Brothers., watch manufacturers, of New York City. Born at Delta, Eaton County, Michigan, October 29, 1865, a son of Orville Boudinot and Mary Elizabeth (Beers) Ingersoll. At the early age of fifteen years he left home and went to New York City, where he entered the employ of his brother, Robert H., who was then engaged in the business of manufacturing rubber stamps. Since 1880 he has been continuously associated with his brother in various business enterprises and in the direction and management of the Ingersoll organization. Married Eleanor Ramsey Bond of Brooklyn, New York, July 5, 1898. Residence, South Orange, New Jersey.

Ingersoll, Robert Hawley Founder and President of Robt. H. Ingersoll and Brothers., watch manufacturers, of New York City. Born December 26, 1859, of Orville Boudinot and Mary Elizabeth (Beers) Ingersoll, at Delta, Eaton County, Michigan, he received his early education in the public schools of his native town. In 1879, at the age of nineteen years, he came to New York City, and in the following year engaged in the business of manufacturing rubber stamps; later, he established a mail order business, selling various "dollar" specialties and novelties. While engaged in this business he conceived the idea and in 1892 commenced the manufacture of the "dollar watch," since which time over 50,000,000 watches have been produced and sold by the Ingersoll organization. Married June 20, 1904, to Roberta Marie Bannister of Green Bay, Wisconsin. Residence, Oyster Bay, Long Island.

Ingersoll, William Harrison Marketing Manager of Robert H. Ingersoll and Brothers., watch manufacturers, New York City. Born March 22, 1879, near Lansing, Michigan. He received a grammar and high school education and three years' technical training for electrical engineer. In 1901 he entered business in the retail sporting goods store of Robt. H. Ingersoll and Brothers. in New York City and was soon placed in charge of the Ingersoll watch advertising, over which he exercised close supervision ever since, except for two periods prior to 1908, when he sought and gained valuable outside experience in other capacities, such as salesman and as manager of the Ingersoll watch business in Canada; he then became advertising manager, later sales and advertising manager and then general marketing manager for developing all markets of all countries of the world for the Ingersoll products. Active in the promotion of advertising research, Mr. Ingersoll was one of the founders of Truth in Advertising work, assisted in establishing a Fellowship in Advertising Research at Columbia University, New York City, and has written and lectured extensively on salesmanship, advertising, marketing and related subjects. His residence is at Maplewood, New Jersey.

Ingold, Franz A Swiss watchmaker who had the idea of making watch parts on the interchangeable plan long before it was put into practice anywhere. He was ill-received by labor and capital alike when he presented his plans in France, England, and America. In England he was nearly mobbed. In 1842-43 he obtained patents on some machinery in this line, but the machines were clumsy and for the most part impracticable. There has been a tendency to credit Ingold as the source of Dennison's ideas on this subject, though Dennison says he never heard of Ingold until after he had started manufacturing.

Intercalary Introduced or added arbitrarily to a calendar; for example, the 29th day of February is an intercalary day.

Interchangeability America's greatest contribution to watchmaking has been the standardizing of parts and the manufacturing of each of them, exactly alike, in great quantities. So that repairing an American watch is largely a matter of obtaining a new part similar to the damaged one, and simply putting it in place.

Invar An alloy of nickel and steel claimed to be non-magnetizable. Used for certain parts of watches at the time when non-magnetizable watches were desirable. Invar is practically non-expansible when the nickel in it is about 37%.

Isochronism That property of a pendulum or balance spring by virtue of which its vibrations, of whatever length, are all made in exactly equal periods of time.

Jacks; or Jack o' the Clock Figures on the old turret clocks which automatically struck the hours. They preceded dials tho were usually left when the dials were added. There are Jacks on the clock at St. Mary Steps, Exeter; Norwich Cathedral, South Aisle; and St. Dunstan's in Fleet St., among others.

Jacquemarts

Jacquemarts Figures of man and woman which struck the hours on the clock set up by Philip of Burgundy at Dijon, prior to 1370. G. Peignot says they are so named from Jacquemart, a clock maker of Lille, employed by the Duke of Burgundy in 1442. The lack of co-ordination in the dates tends to controvert the claim.

Jerome, Chauncey Originator of the one-day brass clock movement which enormously increased the American clock business and opened a market for American clocks in Europe. Born in Canaan, Connecticut, in 1793. Established the Jerome Clock Company at New Haven, Connecticut. This was the predecessor of The New Haven Clock Company.

Jewelled Fitted with precious stones to diminish wear as distinguished from precious stones for ornament. In the best watches ruby and sapphire are used. In lower grade watches quartz, amethyst and garnet.

Jewels Used in watches as bushings at the ends of pivots and in other places which sustain much wear. They:

1. Provide smooth bearings for the pivots.
2. Obviate corrosion.
3. Reduce the wear from abrasion.

Sapphire is the best of the jewels in use and ruby second. Chrysolite is also used and garnet, tho the latter is too brittle for most service. This use of jewels was invented by Nicolas Facio a Swiss watchmaker about 1705.

Julian Period A period of 7980 years obtained by multiplying 28, 19 and 15 the numbers representing the cycles of the sun and moon, and the Roman Indiction. It will end 3267 A. D., until which time there cannot be two years having the same numbers for three cycles.

Jura Mountains A watchmaking center in Switzerland. The industry grew rapidly following the success of Daniel Jean Richard in 1679. This section is the center of the system of watch-manufacturing most nearly like the American system. See Geneva.

Jurgensen, Jules One of the most famous watchmakers of the 19th century; a son of Urban Jurgensen, born in 1808. He studied physics, mechanics and astronomy in Paris and London and finally settled in Locle, Switzerland, specializing in pocket chronometers, which have become famous as the Jurgensen watches. He died in 1877; and was succeeded by his son, Jules F. U. Jurgensen.

Jurgensen, Urban A Danish mathematician and watchmaker born in 1776. He practiced his trade for a time in Switzerland, worked in Paris under Breguet and Berthoud, and then in London, before returning to Copenhagen to enter into partnership with his father, the court watchmaker. He was made superintendent of all the chronometers of the Danish navy and received several decorations. He died in 1830.

Kew Observatory The central meteorological observatory of the United Kingdom. Established at Richmond in 1842 and afterward transferred to the Royal Society. Since 1900 it has been a department of the National Laboratory. Important to the watch business because of the famous Kew tests of timekeepers and awards for accuracy of performance.

Keyless Watches Watches winding without a key. Such watches were made as early as 1686 but did not come into general use until 1843, when Adrien Phillipe (Geneva) introduced the "shifting clutch" type, and when the "rocking bar" mechanism was introduced in 1855. These are the types in use today. Self-winding watches have been made from time to time. Napoleon is said to have had one which wound automatically from the motion of being carried. The abandonment of the key nullified the usefulness of the fusee, although some keyless fusee movements were attempted.

Knuckles The rounded parts of a watchcase that form the hinges or joints. Usually two on the cover.

La Chaux de Fonds A watchmaking center in Switzerland which, in 1840, with a population of 9678, had 3109 watchmakers. At present it is the leading exporter of gold watches in Switzerland. In this section the system of manufacturing is much like the American system.

Laminated Made up of tin sheets of beaten, rolled or pressed metal. In the compensation balance the sheets are of brass and steel, or brass and aluminum.

Lancaster, Pa. A town where there have been watch factories for upwards of fifty years.

Lange, Adolph An eminent Dresden watchmaker born there in 1815, famous for his astronomical clocks, chronometers, and fine watches. Under the direction and with the assistance of his government he established the extensive watchmaking industry of Glashutte. He died in 1875.

Lantern Pinion A pinion consisting of two circular metal end plates usually of brass joined by short steel wires which act as cogs in a gear.

Latitude 1. In astronomy, the angular elevation of a heavenly body above the ecliptic. 2. In geography a distance measured in degrees, minutes and seconds north or south from the equator. 3. In dial work, the elevation of the pole of the heavens; the angle at which the plane of the horizon is cut by the earth's axis.

Lead The continuous action of a wheel tooth which impels the leaf of a pinion or the pallet of a balance.

Leap-Year See Calendar, Gregorian.

Leaves The name applied to the teeth of a pinion wheel.

Lepaute, J. A. 1709-1789. A French clockmaker famous for his turret clocks; the inventor of the pin-wheel escapement and an authoritative writer on horological subjects. He wrote "Traité d'Horlogerie" which was afterward revised and added to by Lalaude.

Lepire, Jean Antoine Born 1720. Died 1814. A celebrated watchmaker of Paris in the 18th century. About 1770 he introduced bars to take the place of a top plate, omitted the fusee, used a cylinder escapement and supported his mainspring barrel arbor at one end only. He attempted to establish a watch factory for Voltaire at Ferney but with no success. He is sometimes credited with making the first thin watch.

Le Roy, Julien 1686-1759. A French scientist and watchmaker. He invented the horizontal movement for turret clocks, a form of repeating mechanism. He constructed the first compensation balance.

Le Roy, Pierre 1717-1785. Son of Julien Le Roy. Esteemed the greatest of all French horologists. He invented a form of duplex escapement and an escapement which formed the basis for the present chronometer escapement.

Lever That part of a lever escapement to which are attached the pallet arms, and which thus transmits motion from the escape wheel to the balance.

Lift, or Lifting Arc That portion of the oscillation of a balance during which it received its impulse. The remainder of the turn is called the supplementary arc.

Lightfoot, Peter A Glastonbury monk, maker of the Glastonbury and Wimburne clocks, 1335.

Lips In a cylinder escapement, the rounded edges of the cylinder through which the escape wheel gives impulse to the balance.

Locking 1. The stopping of the escape wheel of a watch or clock. 2. The portion of the pallet on which the teeth of the escape wheel drop. 3. The depth to which the escape tooth laps upon the pallet at the moment it leaves the impulse face.

Logan, John Born in Lowell, Mass., 1844. Invented a new method of tempering springs and made superior main and balance springs. He was connected for several years with the Waltham Watch Company, during which time he invented many labor-saving machines. Died 1893.

Longitude The circular distance east or west subtending the angle which two meridional planes make at the axis of the earth, one of them being a standard reference meridian.

Longines A watch factory at St. Imier in the Jura Mountains, near La Chaux de Fonds, established in 1874. Here all parts are made under one roof and the work is done by machinery.

Lower Plate The plate in a watch nearest the dial. Also called the "dial plate." It carries the lower pivots of the movement.

Luitprand A monk of Chartres who revived the art of glass-blowing at the end of the 8th century. To him is sometimes ascribed the invention of the sand-glass.

Luminous Dial A watch dial whose hands and figures are so treated as to be visible in the dark. Formerly accomplished by a phosphorescent paint which required frequent exposure to sunlight to be effective and retained its luminosity only an hour or two. Now effected by means of a compound absolutely independent of the sunlight and of a lasting glow. See Radiolite.

lunette

Lunette The usual form of rounded watch crystal.

Mainspring The long steel ribbon used for driving a clock or watch. The spring is coiled into a circular metal box called the barrel and the outer end of the spring is fastened to the barrel; the inner end to the arbor of the great wheel. First applied, replacing weights, by Peter Henlein of Nuremberg, about 1500.

Maintaining Power The device for driving the train while a watch or clock is being wound.

Marsh, E. A. An important figure in watch manufacturing in America for a number of years. Born at Sunderland, Conn., in 1837, in 1863 he entered the employ of the Waltham Watch Company and rose to the position of General Superintendent. In 1908 he retired from active service but retains his connection with the company as consulting superintendent. Besides his practical services to the watchmaking industry Mr. Marsh wrote "The Evolution of Automat Machinery," in 1896.

Massey, Edward An English watchmaker of the early nineteenth century. He invented the "crank roller" escapement, a kind of keyless winding for watches, and many other watch parts.

Mean Solar Day The average length of all the solar days in a year. This period is divided into 24 parts, or hours.

Mean Time Clocks, watches, etc., are made to measure equal units of time instead of the apparent time indicated by the sun. Mean time and true solar time agree only four times in a year. See Equation of Time.

Mercer's Balance-A balance of the ordinary kind fitted with an auxiliary a laminated arm of brass and steel fixed at one end to the central bar of the balance and on its free end carrying two adjustable screws. This auxiliary may be arranged for either extreme of temperature with great accuracy.

meridian dial

Meridian Dial A dial for determining when the sun is on the meridian. It is very simply constructed. For directions see "Watch and Clockmakers' Handbook," by F. J. Britten.

Meridian Watch A watch which shows the time in a number of places in different parts of the world. It is set to Greenwich time and marks the difference between this and the time of all the great metropolitan cities in both hemispheres.

metronome

Metronome An instrument for indicating and marking exact time music. It consists of a counterbalanced, or reversed, pendulum, which may be regulated to swing at any desired number of vibrations per minute.

Middle Temperature Error The compensation balance does not exactly meet the temperature error. The rim expands too much with decrease of temperature and contracts too little with the increase. Hence a watch or chronometer can be correctly adjusted for two points only. The unavoidable error between is the middle temperature error.

Minute The sixtieth part of a mean solar hour.

Minute Hand The hand on a clock or watch which indicates the minutes. In the earlier days clocks had no minute hand. It was first concentered with the hour hand in 1673.

Minute Wheel The wheel which carries the minute hand and is driven by the cannon pinion.

Minute Wheel Pin or Stud The stud fixed to the plate on which the minute wheel pinion turns.

Minute Wheel Pinion or "Nut" The pinion in watches on which the minute wheel is mounted and which drives the hour wheel.

Moment of Inertia The resistance of a body in motion (or at rest) to a change in the velocity or direction of its motion. In a rotating body the sum of the products formed by multiplying the mass of each particle by the square of its distance from an axis.

Month An arbitrary division of the year, varying in the number of days it contains, according to the calendar in use. See Calendar.

Mortise A slot or hole into which a tenon of corresponding shape is to be fitted.

Moseley, C. S. A pioneer in the field of designing and building automatic watchmaking machinery. He invented some of the most delicate and complicated tools and mechanisms used in watch manufacture. He was early connected with the Waltham Co., master mechanic for the Nashua Co., during its brief history; and later general superintendent of the Elgin National Watch Company.

Motion The wheels that carry the hands: cannon pinion, horn wheel and minute wheel and pinion.

Motion Work The wheels in a watch which make the motion of the hour hand one twelfth as rapid as that of the minute hand.

Movement The watch or clock complete, without dial or case the mechanism of the watch or clock.

Mudge, Thomas An English watchmaker of the 18th century. Born at Exeter in 1716, died 1794. In 1793 he received from Parliament three thousand pounds as a recompense for his improvements in chronometers. His work was celebrated for its excellence.

Name Bar The bar which carries the upper end of the arbor of a watch barrel.

Naval Observatory The United States Naval Observatory at Washington, D. C. There is there a superlatively accurate clock from which the time is flashed electrically to all parts of the United States.

Neuchatel A town in the Jura Mountains' watch manufacturing district of Switzerland. A Cantonal Observatory at Neuchatel helps establish the reputation for the accuracy of Swiss watches.

non-magnetic watch

Non-Magnetic Watch A watch in which the quick-moving parts lever, pallets, balance spring, etc., are made of some other metal besides steel as aluminum bronze, invar, etc.

Nuremberg A German city where Peter Henlein made the first watch. It was one of the chief clock centers of the 16th and 17th centuries and with Augsburg and Ulm supplied the markets of Europe with the first small clocks.

Nuremberg Eggs Watches made in Nuremberg in the shape of eggs. If not the first watches at least very early examples.

Obelisk A square shaft with a pyramidal top. The ancient Egyptian obelisks are thought to have served as gnomons.

ogive

Ogive A pointed arch of the architectural type known as Gothic.

Oil Sink The cavity around the pivot hole in watch and clock plates, designed to hold a small particle of oil in contact with the pivot.

Ormolu Gilt or bronzed metallic ware, or a fine bronze which has the appearance of being gilded. Used for ornamenting the cases of fine old clocks.

Orologe An obsolete form of horologe. See Horologe.

Orologiers An obsolete form of horologers, a term not now in use but signifying men who constructed time-pieces.

orrery

Orrery A planetarium; an instrument showing the relative motions, positions and masses of the sun and planets. It was so named from Lord Orrery, for whom the first modern planetarium was made in England.

Oscillation The movement back and forward of a pendulum or the swing of a balance spring. The vibration.

Overbanking Pushing of the ruby pin past the lever, caused by excessive vibration of the balance. In a cylinder escapement the turning back of the cylinder until an escape wheel tooth catches and holds it. In a chronometer escapement the second unlocking of the escape wheel from the same cause.

Overcoil The outermost coil of a Breguet spring which is bent back across the coil toward the center.

Pacificus Archdeacon of Verona, died about 850 A.D. It is claimed by some that he made a clock furnished with an escapement. (Bailly.) But this is not proved, and others believe it to have been merely a water-clock.

Pad The pallet of the anchor escapement for clocks.

Pair Case At one time watches were made with two or even three separate cases. The outer one of shagreen tortoise shell, or some other ornamental material was sometimes for the protection of the delicate enamel on the inner case. Sometimes as in the case of repeaters the inner case was pierced to emit the sound. Then the outer one served as dust protection to the works.

Palladium A soft metal formerly used in alloy with copper and silver for the balance and balance spring of non-magnetizable watches. Too soft to be as serviceable as steel, it has been superseded by a platinum alloy.

Pallet Has different meanings, even among watchmakers. Generally, the part through which the escape wheel gives impulse to the balance or pendulum.

Pallet Staff The arbor on which the pallet is mounted, and on which it turns.

Pallet Stone The jewel on the contact face of the pallet, where it is struck by the teeth of the escape wheel.

Parallax The apparent angular displacement of a heavenly body due to a change of the observer's position.

Pedometer An instrument which registers the number of paces walked hence if properly adjusted to the length of step of the wearer it gives the distance traversed.

Pendant The small neck and knob of metal connecting the bow of a watch case with the band of the case.

Pendulum A body suspended by a rod or cord and free to swing to and fro; used in clocks to regulate the velocity with which the driving power moves the wheels and hence the hands. The isochronism of a pendulum vibrating in a cycloidal arc was first discovered by Galileo but he did not apply it to clocks. Most authorities credit Christian Huyghens with that adaptation to instruments for keeping time. The pendulum was first suspended by a silk cord and thus vibrated in a circular instead of cycloidal arc. "Huyghens' Checks" were an unsuccessful attempt to remedy this. Dr. Hooke succeeded in remedying it by suspending the pendulum by a flat ribbon of spring steel.

gridiorn pendulum

Pendulum, Gridiron Invented by Harrison in 1726, and still with slight improvements an effective timekeeper. The rod of this pendulum is constructed of five steel and four brass rods so arranged that those which expand most are counteracted by those of less expansion, and the length of the pendulum remains constant.

Pendulum, Mercurial Compensation A pendulum having for a bob a jar of mercury which expands upward with the increase of temperature thus counteracting the lengthening of the rod from the same cause. Invented by Graham about 1720. With slight improvements still in use and keeps time very accurately.

torsion pendulum

Pendulum, Torsion A pendulum vibrating by the alternate twisting and untwisting of an elastic suspension. The body is a horizontal disc weighted around its edges, and its suspension a steel or brass wire. The period of a torsion pendulum being much longer than a vibrating pendulum of the same length, the time of running is longer. Clocks fitted with torsion pendulums have run a year on one winding.

Pendulum Swing The short ribbon of spring steel which suspends the pendulum of a clock.

Penetration of Gearing The depth of intermeshing of the teeth of pinion and wheel.

Phillips Spring A balance spring with terminal curves after rules laid down by M. Phillips, an eminent French mathematician. A term seldom used though his curves are generally followed.

pillar

Pillar The three or four short brass posts which keep the plates at their proper distance apart. In early days made in very artistic and elaborate shapes. Later they became plain straight cylindrical columns.

Pillar Model A type of movement in which the works are hung between two plates supported and separated by posts or pillars and forming all the principal bearings of the movement. Only average adjustment is possible in this model. In this model the plate is sometimes cut away to imitate a "bridge model." The opposite extreme in construction to the "bridge model."

Pillar Plate The lower plate of a watch movement the one nearest the dial to which the pillars are solidly fixed, in a "pillar model."

Pinchbeck, or "Pinchbeck Gold" An alloy of three parts zinc to four of copper which "resembles gold in color, smell and ductility." So called from its inventor Christopher Pinchbeck (1670-1732) who during his life guarded the secret of its composition very jealously.

Pinion The smaller of two toothed wheels that work together. The teeth of a pinion are called leaves. See also Lantern Pinion.

lantern pinion

Pinion, Lantern A pinion consisting of two circular metal plates joined by short steel wires.

Pitch The length of the arc of the circumference of the pitch circle from center to center of two adjacent teeth.

Pitch Circle The geometrical circle traced with the center of the wheel as its center and at which the curved tips of the teeth begin. The diameter is proportional to the number of teeth determined upon. The proportion of the pitch circles of a wheel and a pinion gearing together is determined by the ratio of revolutions desired.

Pitkin, Henry With his brother, James F., he started at Hartford, Conn., in 1838, the first factory for machine-made watches in the United States. They made their own machinery, which was very crude. After making about 800 watches they were forced to abandon the project, being unable to compete with cheap foreign watches. He died in 1845.

Pivots The ends of the rotating arbors in a watch that run in bearings.

Planetarium An astronomical clock which exhibits the relative motions and positions of the members of the solar system. Has no regulating system and usually no driving power but is run by turning a crank by hand.

Plates In watches and small clocks the circular discs of brass to which the mechanism of the watch is supported. In large clocks the plates are usually square-cornered oblong. See Pillar Plate, Top Plate, Half Plate, Full Plate, etc. In half-plate, and three-quarter-plate types of watches part of the disc is cut away.

Pocket Chronometer A watch with a chronometer escapement.

Polos A basin in the center of which the perpendicular staff or gnomon was erected, and marked by lines for the twelve portions of the sun-lit day. Herodotus ascribes its invention to the Babylonians, Phavorinus claims it for Anaximander and Pliny for Anaximenes. Also called "Heliotropion."

Potance or Potence A vertical or hang down bracket, supporting the lower end of the balance staff in full-plate watches.

Prescot A town in a remote part of Lancashire for years the center of the movement trade in England.

Push Piece 1. The milled knob pushed in from the pendant to open the case. 2. The boss pushed in when the watch is to be set.

Quare, Daniel 1649-1724 Claimed the invention of the repeater, and backed by the Clockmakers' Company obtained the patent against Barlow from James II. Also credited with the invention of equation clocks. He was master of the Clockmakers' Company in 1708. He first used the concentred minute hand in England, but Huyghens had preceded him in this in the Netherlands.

Quarter 1. A term in common use for the period of three months a quarter of the year. 2. The fourth part of an hour 15 minutes.

Quick Train A watch movement balance vibrates 18,000 times per hour.

Unequal mainspring pull is less felt in the quick train. Used generally in Switzerland and America, and a feature of practically all modern watches.

Rack A straight bar, or segment of a circle, with teeth along one edge. It has a reciprocating motion.

"Radiolite" Trade name adopted by Robt. H. Ingersoll and Brothers. for their watches having black faced dials with luminous hands and numerals. Composed of a substance in which genuine radium is used in minute proportions.

Radius of Gyration The distance from the center of gyration to the axis of rotation.

Ramsey, Davis One of the earliest British watchmakers of renown. He was appointed "keeper of clocks and watches" to James I, and appears to have retained his appointments after the death of the latter. He was the first master of the Clockmakers' Company tho he seems to have taken little active part in the management thereof. Scott introduces him into his story "The Fortunes of Nigel" as a Keeper of a shop a few yards east of Temple Bar. Without doubt he was the leading clockmaker of his day. He died in 1655.

Ratchet The pawl, or dog, which engages in the teeth of a ratchet wheel and prevents it from turning backward. It is held lightly against the periphery of the ratchet wheel by a small spring known as the ratchet spring.

ratchet wheel

Ratchet Wheel A wheel with triangular teeth fixed on to an arbor to prevent the latter from turning backward. The fronts of the teeth are radial, the backs straight lines running from the tip of one tooth to the base of the next. In going-barrel, keyless watches the ratchet has epicycloidal teeth. By "the ratchet" in a watch, chronometer or clock with mainspring is meant the ratchet fastened to the barrel arbor to prevent the mainspring from slipping back when it is being wound.

Recoil In recoil escapements the pallets not only stop the escape wheel but actually turn it backward a slight distance. This backward motion is called the recoil.

Regulator 1. A standard clock with compensated pendulum with which less accurate movements are compared. 2. The lever in a watch by which the curb-pins regulating the swing of the hairspring are shifted.

Remontoire An arrangement in the upper part of the going train by which a weak spring is wound up or a small weight is lifted that gives impulse to the escape wheel at short intervals. Its use is to counteract the irregularities in impulse due to the coarse train, etc. They are delicate and complicated and now superseded by the Double Three-legged Gravity Escapement.

Repeater A striking watch or clock which by the pulling of a string or the pressing of a button could be made to repeat the last hour and part hour, struck. In vogue during the 18th century. Credit for the invention was disputed by Daniel Quare and Edward Barlow. James II gave the decision in favor of Quare whose mechanism was a trifle simpler.

Repousse A kind of chasing in which the metal is punched or pressed from the back bringing the design into higher relief than by the usual method of indenting.

portable sun-dial

Ring-Dial See Sun-dial, Portable.

Richard, Daniel Jean A Swiss watchmaker, born at La Sagne in 1665. At fifteen a watch having come into his hands, he constructed a similar one unaided. That was the first watch made in Neuchatel. After a time in Geneva he set up business in La Sagne, afterwards moving to Locle. He created the watch industry of Neuchatel and saw it grow to a neighborhood of five hundred workers. He died at Locle 1741. In 1888 a bronze statue was erected to him there.

Robbins, Royal E. Born in Connecticut 1824. He was essentially one of the "fathers" of American watchmaking because it was through his financing and clever management that the first watch company finally succeeded in making a financial success.

Roller The circular plate in a lever escapement, into which the ruby pin is set.

Roller-Jewel Same as "impulse pin."

Roman Indiction A period of fifteen years appointed by the Emperor Constantine 312 A. D. for the payment of certain taxes.

Rose Engine A lathe in which the rotary movement of the mandrel is combined with a lateral, reciprocating movement of the tool rest; used for ornamenting the outside cases of watches with involved curved engraving.

Ruby Pin The impulse pin in a lever escapement, made of a ruby.

Ruby Roller The roller in a duplex escapement against which the teeth of the escape wheel are locked.

Run In the lever escapement, the extent of the movement of the lever toward the banking pins after the "drop" on to the locking.

Sabinianus Pope from 604 to 606. Said to have invented a clock in 612 A. D., but the clock he is supposed to have built was probably only another of many forms of clepsydrae, or water clocks.

Safety Pinion A center pinion in a going-barrel watch which allows the recoil of the barrel if the mainspring breaks.

sand-glass

Sand-Glass (Clepsammia) A dumb-bell-shaped glass globe containing sand, and with a small aperture through which the sand flows in a certain fixed time. The most common form is the hour-glass but many others are in use as the three-minute glass for boiling eggs, the two-minute glass used by the British Parliament, etc. Dried and finely powdered eggshell sometimes used in place of sand. The principle is the same as that of the simplest form of clepsydra. See Hour-Glass.

Sandoz and Trot A firm which established the first watch factory in Switzerland in 1804. Previous to that time watchmaking had been a house industry.

Second One-sixtieth of a minute: 1-3600 of a mean solar hour.

Secondary Compensation Same as "auxiliary compensation." See Auxiliary.

Seconds Hand The hand on the dial of a clock or watch which revolves once a minute. Sometimes small and set in a small circle of its own. Sometimes long and traverses the whole dial. See Center-seconds and Sweep-seconds.

Seconds Pivot The prolongation of the fourth wheel arbor to which the seconds hand of a watch is fixed.

Seconds, Split Divided seconds into quarters, or fifths; measured by a chronograph.

Shadow A darkened space resulting from the interception of light by an opaque body.

Shagreen Made from the tough skin that covers the crupper of a horse or ass. Rough seeds are trodden into the skin and then allowed to dry. The seeds are shaken out and the skin dyed green. Then the rough surface is rubbed down smooth leaving white spots on the green ground. Also made from the rough skin of sharks and dolphins. Formerly used a great deal for the outer cases of watches. See Pair Cases.

Sherwood, Napoleon Bonaparte Born in 1823. About 1855 he entered the watchmaking business in the employ of the Waltham Watch Co. He revolutionized jeweling methods and invented among other things a "Counter-sinker," "End-shake tools," "Truing-up tools" and "Opener." In 1864 he organized the Newark Watch Company but within a few months severed his connection with it. He died in 1872.

Sidereal Time The standard used by astronomers; measured by the diurnal rotation of the earth, which turns on its axis in 23 hours, 56 minutes, 4.1 seconds. The sidereal day is therefore 3 minutes, 56 seconds shorter than the mean solar day. Mean time clocks can be regulated with greater facility by the stars than by the sun for the motion of the earth with regard to the fixed stars is uniform. Clocks all over the United States are so regulated from the Naval Observatory at Washington.

Side-Shake Freedom of pivots to move sideways. See End-Shake.

Slow Train A train whose balance vibrates 14,400 times an hour. Now never used in pocket watches because of susceptibility to inequalities in the pull of the mainspring, jars, sudden movements, etc. Used, however, in marine chronometers.

Snail A cam shaped like a snail, used generally for gradually lifting and suddenly discharging a lever, as in the striking mechanism of clocks.

Snailing A method of ornamenting with circles and bars parts of a watch movement which it is not desirable to polish highly.

Solar Time Time marked by the diurnal revolution of the earth with regard to the sun, of which the midday is the instant at which the sun appears at its greatest height above the horizon. This instant varies from twelve o'clock mean time because the earth also advances in its orbit and its meridians are not perpendicular to the ecliptic.

spandrels

Spandrels The corners of a square face outside the dial of a clock. Formerly very beautifully decorated. The age of the clock can be told approximately from the form of ornamentation employed.

Split Seconds A chronograph in which there are two center-seconds hands one under the other which can be stopped independently of one another.

Spring-Clocks Clocks whose driving power is a coiled spring instead of a weight.

stackfreed

Stackfreed The derivation of the word is obscure; it is possibly Persian. A device to counteract the difference in power of the mainspring at the different stages of its unwinding. Fixed to the mainspring arbor above the top plate is a pinion having eight leaves, which gears with a wheel having twenty-four teeth, which do not quite fill out the circumference of the wheel. Fastened to the wheel is a cam, concentric for about seven-eighths of its circumference and indented for the remainder. Into a groove in the concentric portion of the edge is pressed a roller which is pivoted at the free end of a strong curved spring. When the mainspring is fully wound the roller rests in the curved depression of the cam and the effort required to lift the roller up the incline absorbs some of the mainspring's power. On the other hand when the mainspring is nearly run down, the roller is descending an inclined plane and absorbs less of the power. Not an acceptable device and now rarely met with.

Stem-Winding The ordinary method of winding keyless watches by means of a stem running through the pendant.

Stop Work An arrangement for preventing the overwinding of a mainspring or a clock weight.

Stratton, N. P. One of the early watchmakers connected with American manufacture. He was an apprentice of the Pitkin Bros., and was sent by the Waltham Company to England in 1852 to learn gilding and etching. He was made assistant superintendent of the Waltham Co. in 1857. He invented a mainspring barrel and a hair-spring stud which were later adopted by the Waltham Company.

Striking-Work The part of a clock's mechanism devoted to striking. The chief forms are Rack, and Locking-plate, or Count-wheel. See separate articles.

Striking-Work, Locking-Plate, or Count-Wheel Used in turret clocks where there is no occasion for the repeating movement. This form of striking work does not allow of the repetition or omission of the striking of any hour without making the next one wrong.

Striking-Work Rack A form of striking work used largely in house clocks; the number of blows to be struck depends merely on the position of a wheel attached to the going part. In this form the striking of any horn may be omitted or repeated without deranging the following strikes.

Stud 1. A small piece of metal pierced to receive the outer or upper coil of a balance spring. 2. The holder of the fusee stop-work. 3. Any fixed holder used in a watch or clock, not otherwise named, is called a stud.

Style The finger or gnomon on a sun-dial whose shadow, falling on the plate, indicates the time.

Sully, Henry An English watchmaker of the early eighteenth century who lived most of his life in France. He presented the French Academy with a marine timekeeper superior to the timepieces of the period, and a memoir describing it. He died shortly afterward and advance in the art was delayed.

Sun-Dial A device for telling time by the shadow of a style, cast by the sun, as thrown upon a disk or plate marked with the hour lines. Dials were named from their positions equinoctial or equatorial; east; erect or vertical; horizontal; inclining, etc., or from their purpose or method of use, as portable, reflecting, etc., or as in the case of the ring-dial, from their form. The word is derived from the Latin dies. The style in the earliest dials was a vertical staff, but later it was found that reasonable accuracy could only be obtained by a style set parallel to the earth's axis that is, inclined to the horizontal at the angle of latitude of the locality in which the dial was set.

Even before the first astronomical discoveries of the Babylonians, people had felt some need of a convenient device to mark and measure the passing of the time, especially the shorter divisions of recurring time, the time of day. Sunrise and sunset marked themselves by the horizon, but noon was harder to determine, and the points of mid-morning and mid-afternoon harder still. And with the knowledge of those regular movements in the heavens which determine time on earth, and with the closer division of the day into its hours, that need became a sheer necessity.

The obvious measure of the sun's movements was the moving shadow cast by the sun itself. And the earliest device for recording time was naturally the sun-dial. Its origin fades into the twilight of antiquity. Long before we know anything about him, primitive man measured the moving shadow of some tree. And it occurred to him to set up a post or pillar in some convenient place, and mark out the positions into which the shadow swung. The earliest sun-dials were of this pattern, with a vertical pointer of gnomon, and the hours marked upon the ground. And it is related of the early Greeks that they told the time individually by marking and measuring the length of their own shadows. But the measure of time by the length of a shadow is very irregular at best, because of the yearly motion of the sun. The shortest shadow of the day will indeed fall at noon. But that noon shadow will vary in length according as the sun's noon is high in Summer or low in Winter; and so the whole scale of lengths will be different for every day in the year. If a three foot shadow means mid-afternoon today, it will mean quite another time tomorrow. And for measuring by the direction of the shadow, the vertical gnomon is more irregular still. For the swing of the shadow would depend not only upon the sun's motion across the sky from East to West, but also upon his slant North and South along the sky. And this would change from day to day. The difficulty was to make a dial of which the shadow would move as regularly as the sun moves.

ANCIENT GREEK HEMICYCLE
ANCIENT GREEK HEMICYCLE

This the ancients accomplished in a very simple and ingenious way. The sun moves in the sky as it were upon the inner surface of a hollow globe or sphere. So they made the dial a little hemisphere, place with its hollow side up toward the sky as a bowl stands on a table. The pointer was placed above and to the South of this, on the side toward the sun; and the Time was marked by the shadow of the tip end of the pointer which was a little ball or bead. The path of this shadow across the bowl reproduced exactly on a small scale the path of the sun across the great bowl of the heavens. And it was then an easy matter to mark off the bowl into equal divisions which the shadow would cross at equal intervals of the day. Of course, the track of the shadow changed with the season of the year. But it moved always as the sun moved, and just as regularly, giving a true measure of the solar day.

The principle of this was applied in several interesting variations. The defect of the Hemicycle, as this hollow type of dial was called, was that it could not be read accurately for short intervals. A shadow moving only a few inches in the whole day must move so slowly that one could hardly see it move at all. To mark the minutes, it must move faster, just as the minute hand of your watch moves faster than the hour hand, and the second hand faster still. One cannot read seconds from the hour hand, however accurately it moves, because it moves so slowly. So the idea was applied by making the shadow move across a street or courtyard, down one side and across and up the other side, as the sun opposite went up and across and down the sky. Sometimes the place was partly roofed over, and a single beam of light admitted through a small hole at the South end. The resulting spot of light would then move in the same way. The long sunbeam or shadow moved faster, and so could be read at shorter intervals. The Hemicycle is not certainly known to have been invented until long after this, about B. C. 350. But the principle of it is so simple and so entirely such as would occur to an intelligent man still ignorant of its mathematical explanation, that we may not unreasonably suppose it to have been discovered by experiments long before.

ANCIENT ROMAN HEMICYCLE
ANCIENT ROMAN HEMICYCLE

The final improvement of the sundial was the discovery that by slanting the gnomon so that it pointed exactly toward the North Pole of the sky, the direction of its shadow could be made to show the solar time correctly. Since the sky is infinitely far away, the line of the gnomon would then lie parallel to the axis of the heavens. And the sun, moving parallel to the celestial Equator, would always move straight across the gnomon. In other words, he would practically revolve around its sloping edge. Therefore the North and South motion of the sun would be as it were along the edge of the gnomon, and would not influence the direction of the shadow at all. His East and West motion alone would govern the swing of the shadow; and the dial would keep true time with the sun for every day in the year. There was no longer any necessity for hollowing out the dial itself into the concave form; it might just as well be the more convenient flat surface, and this might be either vertical or horizontal, so long as the gnomon pointed straight to the Celestial Pole. All that was needed was to mark out on the dial the true direction in which the shadow fell for each hour of the day.

OLD ENGLISH DIAL
OLD ENGLISH DIAL

Just when or by whom the instrument was thus scientifically perfected is not known. The calculations necessary to the projection of the hour lines upon a flat surface could hardly have been performed before Greek times. The Greeks ascribed the invention of the sundial to Anaximander, in the sixth century B. C., but sundials of various types had been known in various parts of the world long before then. On the other hand, the Hemicycle remained the common form of the instrument all through the classic period and even afterwards. The Babylonians were quite capable of understanding the principle of the sloping gnomon. And once this was discovered, it would have been entirely practical to set up the new dial beside a Hemicycle or Clepsydra, and find the angles of the hour lines by experiment. These, once laid out correctly, would be determined once for all. Even at its best the sundial had certain very marked limitations. Scientifically constructed, it would keep accurate time according to the visible sun. But it could not be read accurately unless made inconveniently large. It was inaccurate when removed from its original latitude, or displaced from a true North and South position; so that in any portable form it became a very rough measure indeed. Moreover, it was of course entirely useless at night or in bad weather or in shadow. And finally, it was never absolutely exact under the most ideal conditions, because of what is known as the Equation of Time. The Earth does not, in fact, move around the sun at an absolutely regular rate of speed; it moves a trifle faster during certain parts of the year and slower at others. The sun therefore varies correspondingly his apparent speed along the Ecliptic, so that even from noon to noon the sun is not always precisely on time. He may be as much as fifteen minutes late or early, according to the season. And our modern days are measured according to the sun's average rate, so as to allow for this variation and keep every day exactly twenty-four hours long. This of course no sun-dial can possibly be made to do, since it must follow the actual sun.

The sun-dial has remained in use to the present day. It seems strange to think of a sun-dial being used as a standard for setting clocks and actually to regulate the running of trains. But these things were done in civilized Europe within the last half century. It was only when the railroad and the telegraph had made standard time at once necessary and easy to obtain that the sun-dial altogether lost its position of authority.

Sun-Dials, Descriptions Classical sun-dials were of many forms. Vitruvius, the Roman engineer, mentions thirteen, some of them portable; and ascribes the invention of the Hemicycle to the Babylonian astronomer and priest, Berosus. There was a famous dial of this type at the base of Cleopatra's Needle in Egypt. It is now at the British Museum. And the Emperor Augustus, returning from his Egyptian wars, brought home to Rome an obelisk which he set up as the gnomon of a huge dial in the Campus Martius. At Athens there was the famous Tower of the Winds; octagonal in shape, with a weather vane above, and below around the tower, the hours and the winds, to each of which the Greeks gave a personality and a name. There is a curious bit of accidental poetry in the marking of the sun-dial in Greece. The Greek numerals, like the Roman, were simply the letters of their alphabet arranged in a certain order. The hot hours of the day from noon to four o'clock were those commonly devoted by the Greeks to rest and recreation. Reckoning the day from sunrise, this period ran from the sixth hour through the ninth. And the numeral letters for Six, Seven, Eight and Nine, which marked those hours upon the dial,

spell out the Greek word ZʹHOI, the imperative of the verb to live. The poet Lucian thus points the moral:

Six hours to labor, four to leisure give;
In them so say the dialled hours LIVE.

The shepherds of the Pyrenees still consult their pocket dials. And the Turk makes a sun-dial of his two hands by holding them up with the tips of the thumbs joined horizontally and the forefingers extended upward; so that the shadow of one forefinger falls toward the other and by its position roughly indicates the time. But even now, when it has nearly gone from practical use, the sun-dial, as an appropriate adornment of our public parks and our private gardens, is becoming increasingly fashionable in our own generation.

OLD FRENCH WALL DIAL
OLD FRENCH WALL DIAL

Sun-dials are common in almost all parts of the world, and not a few of them have in one way or another become famous. The largest is at Jaipur in India, and was erected about 1730. Its gnomon is ninety feet high and one hundred and forty-seven feet long. A flight of stone steps run up the slope of it, and at the top there is a sort of little watch-tower. And the shadow, which falls upon a great stone quadrant instead of upon a flat surface, moves at the rate of two and a half inches a minute. Another great dial is the so-called Calendar Stone of Mexico, which was made by the Aztec priests more than a hundred years before the Spaniards came. It weighs nearly fifty tons, and is not only a sun-dial but a representation of the zodiac and a diagram of the astronomical changes of the year: thus showing that the ancient Mexicans in their own way paralleled the astrology of the Babylonians on the other side of the world. Probably the most expensive and elaborate sun-dial ever built was the one set up in 1669 by King Charles II of England in front of the banqueting house at White Hall in London. It was in the form of a tall pyramid on which were two hundred and seventy-one different dials, giving not only the hour of the day but various astronomical and geographical indications as well. The place called Seven Dials in London takes its name from a tall pillar with sun-dials around its top which used to stand at the junction of seven streets radiating starwise from that spot as a center. The pillar was overthrown in 1773 by a party of vandals digging for buried treasure which they believed to have been hidden beneath its base. Extensive list, descriptions and illustrations, See Book of Sun-dials, Mrs. Alfred Gatty; Sun-dials and Roses, Mrs. Alice Morse Earle.

OLD ENGLISH PILLAR DIAL
OLD ENGLISH PILLAR DIAL

Sun-Dials, Greek 1. Diogenes asserts that the first Greek dial or gnomon was erected by Anaximander of Miletus. It was probably a vertical rod on a horizontal plane. This was two centuries after the Dial of Ahaz. 2. On the "Tower of the Winds" in Athens a dial on each face.

Sun-Dial, Hollow A form of sun-dial invented by the Chaldean Berosus. A hollow hemisphere with a bead at its center, whose shadow indicated the hour of the day.

Sun-Dial, Mottoes On nearly all sun-dials both ancient and modern there is inscribed a motto usually of the moral significance of the passage of time.

Very ancient also, as well as equally common in modern times is the custom of placing upon the sun-dial some appropriate motto expressive of the mystery of Time. There are hundreds of such mottoes, ranging in sentiment from the old Roman one: Horas non numero nisi Serenas. "I number no

hours but the fair ones," to the couplet of a modern poet:
"Time flies, you say? Ah no,
Alas! Time stays; we go."
And these two thoughts, expressed in many forms, represent fairly the tenor of most of them. There is a story of a lazy apprentice asking a motto for his dial, to whom his master sharply replied: "Begone about your business!" and the fellow, appropriately enough, took that for the motto required. It is at least a familiar sentiment, especially in Puritan times; and equally so during the Middle Ages is that more mystic suggestion, Umbra Dei "the Shadow of God."

portable sun-dial
Sun-Dial, Portable Made in different shapes and upon different plans small enough to carry about. The most common form was the ring dial, consisting of a metal ring with a hole in it through which the light fell upon an inside ring adjustable to the day and month. It required careful orienting to be dependable as a time-indicator.

Sun-Dials, Roman The first dial in Rome was set up B. C. 293 near the temple of Quirinus by Papirius Cursor. It served ninety-nine years; then one more accurate was set up beside it. Before that, no time was noted except the rising and setting of the sun. Emperor Augustus erected a dial at Campus Martius. A dial captured in Sicily during the first Punic war was set up in the Forum about 263 B. C. and used for years before they learned that it was inaccurate in that latitude, being designed for the latitude of Sicily.

Sunk-Seconds A dial in which the seconds circle is sunk below the rest of the dial. It allows the hour hand to be placed closer to the face thus making a thinner model possible.

Supplementary Arc See: "Lifting Arc."

Sweep-Seconds See: Center-Seconds.

Table Roller The roller of a lever escapement which carries the impulse pin.

Tell-Tale Clock A clock by which a record is left of periodical visits of some one as a night-watchman.

Template or Timplet One of the four facets that surround a cut gem.

Tenon A projection at the end of a piece cut to fit into a corresponding mortise.

Terry, Eli The first man to make clocks by machinery in America. When it was learned that he planned to make two hundred clocks he was much laughed at. He was born at East Windsor, Conn., in 1772. His first clocks were made by hand, the movements being of wood. He was the leading maker of wooden clocks in America. He invented the shelf clock which contained distinctly new inventions and he introduced the pillar scroll-top case. He was a mechanical genius and contributed a great deal to developing clock-making in America into a great industry. He died in 1852.

Third Wheel The wheel in the train between the center wheel and the fourth wheel.

Thales A celebrated Ionian astronomer, one of the Seven Sages of Greece. He was born about 640 B. C., and is credited by Herodotus with having predicted an eclipse of the sun occurring about 609 B. C. He was the author of several solutions of geometrical problems. He died about 550 B. C.

Thomas, Seth Born at Wolcott, Conn., 1785. A very successful clockmaker who contributed probably more than any other man toward popularizing the modern cheap clock. The Seth Thomas Clock Co., of today, he started in 1813 with twenty operatives. By 1853 it had nine hundred. He died in 1859.

Three-Quarter Plate A three-quarter plate watch is one in which there is a piece cut out from the top plate large enough to permit the balance to rotate on a level with that plate. It is the most common form at present in use in both cheap and high grade watches, and found in both "pillar" and "bridge" models.

time candle

Time-Candles Candles in alternate black and white sections were used to mark the passage of time in Europe and Asia for a long time. In England and France they were used to limit the bidding at an auction. The phrase "by inch of candle" meant that the one bidding when the flame expired was the successful bidder. King Alfred is said to have used time-candles and to have inclosed them in thin horn plates to protect them from drafts, thus originating the lantern.

Timekeeper Any device primarily concerned with measuring and indicating the sub-divisions of the day.

Tompion, Thomas "The father of English Watchmaking." Born 1638. He was the leading watchmaker at the court of Charles II. He found the construction of the time-keeping part of watches in a very indifferent condition and he left English clocks and watches the finest in the world, although many great improvements were made after his time. He associated closely with such scientists as Hooke, and Barlow, and made practical application of their theories two notable instances being the cylinder escapement and the balance-spring. Tompion was the first to number his watches consecutively for the purpose of identification though he did not so mark his early ones. There is a famous clock in the pumproom at Bath, England, of Tompion's construction. Little is known of his domestic life but he appears to have been unmarried. He died in 1713 and is buried in Westminster Abbey. Tompion was master of the Worshipful Clockmakers' Company in 1704.

Top Plate The plate in a watch farthest from the dial. In full plate watches it is circular; in three-quarter plate or half-plate watches a part is cut away.

Tower of the Winds

Tower of the Winds An octagonal tower north of the Acropolis of Athens spoken of as horological by Vario and Vitruvius. Believed to have had a sundial on each of its eight faces and to have contained a clepsydra fed by a spring.

Train The toothed wheels of a watch or clock which connect the barrel or fusee with the escapement. In a going-barrel watch the teeth about the barrel drive the center pinion which drives the center wheel and then in turn the third wheel pinion, third wheel, fourth wheel pinion and fourth wheel, escape pinion and escape wheel.

Tripping The running past the pallet's locking face, of an escape wheel tooth.

Vacheron and Constantin In 1840 established the first complete watch factory in Switzerland. Not until later, however, was motor power used instead of foot-power; and later still manufacture by machinery. The work in this factory is carried on under a combination of all accepted methods.

Vailly, Dom A Benedictine monk of about 1690 who made a water clock which Beckmann says was the first to be constructed on a really scientific principle. See Clocks, Interesting Old Vailly's.

Van der Woerd, Charles A prominent man in connection with watch manufacturing in this country. In 1864 he invented an automatic pinion cutter; in 1874 an automatic screw machine. From 1876-1883 he was superintendent of the Waltham factory.

Verge The pallet axis of the verge escapement. See diagram of Verge Escapement. It carries the balance at its top.

Verge Watch A watch with a verge escapement.

Vick, Henry de. See De Vick.

Volute A flat spiral.

Volute-Spring A flat metallic spring coiled in a spiral conical form and compressible in the direction of its axis.

Wallingford, Richard An English mechanic and astronomer of the fourteenth century. He made a clock which is supposed to have been the first that was regulated by a fly-wheel. Several authorities, however, claim that Wallingford's "clock" was actually a planetarium.

Waltham A town in Massachusetts the site of the first successful watch factory in America. At present a great watch making center.

Watch In modern parlance, a small timepiece to carry about on the person. Formerly a timepiece which showed time in distinction to clock which struck time. Derham (1734) uses the term to indicate all timepieces driven by springs. The term may have been derived from the Swedish vacht, German wachen, or Saxon woecca. The spaces of time between the fillings of a clepsydra were also called "watches."

Watch Collections For list of principal collections, past and present, see Jewelers' Circular files August to December 1915. List compiled by Major Paul M. Chamberlain of Chicago. For list of principal present collections, see Appendix to this volume derived from the Chamberlain Compilation.

Watchmakers' Schools American. In America these schools usually teach watch-repairing and not the making of watches. Some of them offer courses in making watches but few pupils avail themselves of these courses. List of: De Selins Watch School, Attica, Ind.; Detroit Technical Institute Detroit, Mich.; Kansas City Watchmaking and Engraving School, Kansas City, Mo.; Needles Institute of Watchmaking, Kansas City, Mo.; Bowman Technical School, Lancaster, Pa.; Ries and Armstrong, Macon, Ga.; Drexler School for Watchmaking, Milwaukee, Wis.; Newark Watchmaking School, Newark, N. J.; Philadelphia College of Horology, Philadelphia, Pa.; St. Louis Watchmaking School, St. Louis, Mo.; Schwartzman's Trade Schools, San Francisco, Cal.; Stone School of Watchmaking, St. Paul, Minn.; Waltham Horological School, Waltham, Mass.; Bradley Polytechnic Institute, Peoria, Ill.

Watchmakers' Schools, Switzerland Usually under government management. Teach very thoroughly and completely the art of making a watch from the beginning.

Watch-Papers During the 18th century it was a fad in England and America to carry small round papers, which exactly fitted the case of a watch. On these were portraits and verses, the latter of doubtful merit and usually of sinister or gloomy significance.

Waterbury A town in Connecticut long a center of clock and watch making in America. Home of the original Waterbury watch. Location of principal factory of Robt. H. Ingersoll and Brothers., manufacturers of the Ingersoll watches.

Water-Clock Any device, as a clepsydra, for measuring time by the fall or flow of water. More commonly applied to the type in which wheels are turned by water or in such as those in which water sets machinery of some form in motion as Vailly's water-clock. See Clock, Vailly's.

Wick Timekeeper A wick or rope made of some fiber resembling flax or hemp with knots tied at regular intervals and so treated that upon ignition it would smolder instead of breaking into flame. Early in use in Japan and China. Time was estimated by the burning between the knots.

Wieck, Henry De See De Vick.

Willard, Aaron Born 1757. Probably learned his trade from his older brothers Simon and Benjamin. He made tall, and shelf clocks, later banjo clocks so-called from their shape gallery clocks, and regulators. A better business man than his brothers and successful from the start. His clocks did not lack decorative merit but were inferior to Simon Willard's. He made a greater number than his brother because more successful in a business way.

Willard, Benjamin Older brother of Simon and Aaron Willard. Among the first of American clockmakers. Born 1743. Made, probably, only tall clocks with handsome cases and some with musical attachments. Not so good as the clocks of Aaron and Simon Willard but older and rarer now.

Willard, Simon Born at Grafton, Mass., 1753. One of the earliest Massachusetts clock makers who disputed the claim of the Connecticut makers for the credit of revolutionizing the clock industry in America. So far as cases go they excelled Terry, Thomas, and others. But to the Connecticut makers belongs the credit for having developed clock making into a great industry. Willard at first made eight-day tall clocks and shelf clocks, later wall clocks which he called "time pieces." In 1802 he practically abandoned the making of tall clocks, and confined himself to his "time pieces" and special orders for tower and gallery clocks. For a detailed list of his productions see his Biography by John Ware Willard. He was an intimate friend of Jefferson, Madison and other leading men of the time. Died 1848.

Worshipful Clockmakers' Company of London, The Incorporated August 22, 1631, under special charter by King Charles I of England. Was given the sole privilege of regulating the watch and clock trade in and for ten miles around London.

Webster, Ambrose Mechanical superintendent, and later assistant superintendent, of the Waltham factory until his resignation in 1876. He systematized the work in the shop, standardized the measuring system, and forced automatic machinery to the front. He designed the first watch factory lathe with hard spindles and bearings of the two taper variety. He made the first interchangeable standard for parts of lathes. He invented many machines now in use, among them being the automatic pinion cutter.

Weight-Clock A clock whose driving power is a weight suspended by a cord wound on a drum or cylinder.

Weights The first clocks were made with a weight on a cord which was wound around a cylinder connected with the train. The weight descending caused the cylinder to revolve, setting the train in motion. Too rapid unwinding was prevented by the escapement. The weight as a driving power is still used, especially in large clocks.

Wheel, Count The wheel carrying the locking-plate in a striking mechanism.

Year Astronomically, the period of time occupied by the earth in making one complete revolution around the sun. The calendar year is an arbitrarily determined division of time, approximating more or less closely the astronomical year. See Calendar, Gregorian.

Zech, Jacob Of Prague. Invented the fusee about 1525. The Society of Antiquaries possesses an example of his handiwork a table time-piece with a circular brass-gilt case 9¾" in diameter and 5" high. For minute description see Archaeologia vol. xxxiii.

Zero A time-telling term originating or at least made common during the Great War. Word commonly used in a military sense to indicate a secret instant of time from which an attack in its various stages is scheduled.

Zodiac

Zodiac An imaginary belt 16 degrees in width, spread equally on both sides of the ecliptic (q. v.). It is divided into twelve sections or "signs" which receive their distinguishing names from the twelve principal constellations within the belt. That is how the Babylonians learned to tell the time by looking at the sun and the stars. Only their whole problem was vastly complicated by the daily rotation of the earth on its axis, which of course makes the whole sky seem to turn in the opposite direction day by day. The earth turns in the same direction that it goes round the sun, from West to East. So the heavens turn apparently from East to West, while the annual motion, as we saw just now by the illustration of the clock face, appears in its true direction, Eastward. Also, the great clock of the sky is not from our point of view horizontal, but stood up on edge; and not straight up and down even, but slanted at an angle. So its apparent movements are as it were in several directions at once, and the effect is very confusing. The real motions as they actually do occur are very much simpler and easier to understand. But of these the Babylonians had no idea. They knew only what they could see; and it is all the more wonderful that they contrived to reason out so much and so correctly.

They mapped out a belt or zone around the sky, with the Ecliptic along the middle of it. This they divided into twelve equal parts of thirty degrees each, called Signs or Houses, and each containing a constellation. These constellations were in order, Aries or the Ram; Taurus or the Bull; Gemini or the Twins; Cancer or the Crab; Leo or the Lion; Virgo or the Virgin; Libra or the Scales; Scorpio or the Scorpion; Sagittarius or the Archer; Capricornus or the Goat; Aquarius or the Water-Carrier; and Pisces or the Fishes. We know these by their Latin names, and the whole zone by its Greek name of The Zodiac. But their original titles were much the same, only in a different language. The sun went through one of these constellations each month; and by his position along the Zodiac they told the time of year. Thus the Spring Equinox was where the sun entered the House of the Ram; and that was for the ancients the first day of the new year. The House of the Crab was farthest North, and when the sun got there it was midsummer. The Autumn Equinox was in the House of the Scales; and when the sun reached the House of the Goat, he would be at the Southern or Winter end of his journey. Moreover, since the Moon and the Planets always keep close to the Ecliptic, their apparent motions all lie within the Zodiacal zone. And the Zodiac therefore represented the most important part of the heavens from the standpoint of keeping time; the part, that is, wherein all of those bodies which moved among the stars month by month and day by day appeared to have their motions.